# Moduli, deformations and classifications of compact complex manifolds

D Sundararaman
University of Hyderabad

# Moduli, deformations and classifications of compact complex manifolds

Pitman Advanced Publishing Program
BOSTON · LONDON · MELBOURNE

PITMAN PUBLISHING LIMITED
39 Parker Street, London WC2B 5PB

PITMAN PUBLISHING INC.
1020 Plain Street, Marshfield, Massachusetts

*Associated Companies*
Pitman Publishing Pty Ltd., Melbourne
Pitman Publishing New Zealand Ltd., Wellington
Copp Clark Pitman, Toronto



AMS Subject Classifications: (main) 14J15, 32G05, 32G13
(subsidiary) 14J10, 32G20, 58H05

Library of Congress Cataloging in Publication Data

Sundararaman, D
Moduli, deformations, and classifications of compact complex manifolds.

(Research notes in mathematics ; 45)
Bibliography: p.
1. Complex manifolds. 2. Moduli theory. I. Title.
II. Series.
QA331.S92 515.9'223 80-20825
ISBN 0-273-08458-5



Manufactured in Great Britain

ISBN 0 273 08458 5

# Contents

# Preface

> Riemann's classical problem of Moduli is not a problem with a single aim, but rather a program to obtain maximum information about a whole complex of questions which can be viewed from several different angles.
>
> Lars V. Ahlfors

The aims of this volume are to present a survey of major achievements, of the last two decades, pertaining to the topics mentioned in the title and at the same time to give introductions to the various topics dealt with. Because of these two rather incompatible aims, my presentation has to be unconventional in certain respects. Detailed proofs are omitted; but references are made to where original and simplified proofs (if available) are found. There is an exhaustive bibliography. Many references not cited in the volume are also listed to form a natural sequel. As the above quotation from Ahlfors implies, it is very difficult to give a complete up to date account of the topics. I have tried; I do hope that this volume presents an adequate and satisfactory account.

Readers are assumed to have knowledge of fundamentals of Algebraic Topology, Differentiable and Complex Manifolds and Algebraic Geometry. However necessary background information and motivational remarks are given at appropriate places in order to make the volume accessible to nonspecialists as well.

On various topics of the volume, I had given series of lectures at the Bombay Mathematical Colloquium, the Ramanujan Institute, Madurai University and the International Centre for Theoretical Physics, Trieste, Italy. I take this opportunity to thank Professors K.R. Parthasarathy, T.S. Bhanu Murthy, M.Venkataraman and J.Eells, for their enthusiastic encouragement during these times. At various stages of this work, I had the privilege of receiving critical comments and helpful suggestions from Professors Masatake Kuranishi, Raghavan Narasimhan and David Mumford. I am grateful to them. They are of course not responsible for any inaccuracies in my presentation. I would like to thank M. S. Balasubramani, G. Lakshma Reddy and K. Sithanantham for their help, especially in the compilation of the bibliography.

University of Hyderabad,
Hyderabad,
India.

D. Sundararaman

# 1 The moduli problem



## SECTION 1

The main purpose of this section is to show that the existence of a complex structure is a highly non-trivial phenomenon. In Part A, we consider Topological, PL and smooth manifolds. We state some of the outstanding problems and major achievements in the three closely related theories: Geometric Topology, Algebraic Topology and Differential Topology. We strongly feel that awareness of these problems and achievements, with some familiarity of the important concepts and methods of proof, enhances the proper understanding and appreciation of the important aspects of the main topics dealt with in this volume.

In Part B, we consider the problem of the existence of almost complex structures, complex structures and more generally G-structures and Γ-structures. Even though the main topics of this volume pertain to complex manifolds, it becomes absolutely necessary to consider larger classes : Complex spaces, algebraic varieties, schemes and algebraic spaces. This is because the main problem of compact complex manifolds that we consider admits solutions (if it does) only in the larger classes. Theorems and techniques of these larger classes have often been found to be useful for the study of complex manifolds.

Part C, gives a rapid introduction to the important concepts and theorems of modern algebraic geometry.

## Section 1: Part A: Topological, PL and Differential Manifolds

### I-1-A-(1): Topological manifolds

DEFINITION I.1.1.

A topological m-manifold M is a paracompact Hausdorff connected topological space M such that each point of M has an open neighbourhood homeomorphic to an open set in the m-dimensional real Euclidian space $\mathbb{R}^n$.

Remarks

(1) The uniqueness of the integer m follows for example by applying the Brower theorem on the invariance of domains (for proof see page 95, Hurewicz and Wallman [418].) m is called the dimension of the manifold.

(2) We always take manifolds to be connected. Instead of requiring that the topological space M be paracompact and Hausdorff, it is sometimes convenient to take it to be metrizable (which would imply Hausdorff and Paracompactness).

(3) If in the above definition, we require that each point of M has a neighbourhood homeomorphic to open sets in $R^m_+ = \{x = (x_1, \ldots x_m) \mid x_m \geq 0\}$, we get the concept of a topological manifold with boundary. The points of M that correspond to points in the hyperplane $\{(x_1, \ldots x_m) \in R^m_+ \mid x_m = 0\}$ form the boundary of M, denoted by $\partial M$. It is easy to check that $\partial M$ is an $(m-1)$-manifold (without boundary). Unless otherwise stated, by a manifold we mean a manifold without boundary and by a closed manifold we mean a compact manifold without boundary.

(4) It is clear from the definition, how to define a continuous map from one topological manifold to another. Topological manifolds and continuous maps form a category.

(5) The Recognition Problem: It is a problem to determine which topological spaces are topological manifolds. In dimension 1, two nice results are: (i) If M is a compact connected metric space with exactly two separating points, then M is homeomorphic to the closed interval $[0,1]$. (ii) If M, as in (i), is such that every pair of points separates M then M is homeomorphic to the 1-sphere $S^1$ (for proof see Newman [692]). For details on the Recognition Problem in dimension 2 see Van Kampen [440], and in dimensions $\geq 5$ see Cannon, B.A.M.S 84(1978), 832-866.

### I-1-A-(2): Differentiable Manifolds

Let M be a topological m-manifold. By a differentiable ($C^\infty$) atlas on M we mean a family $(U_i, h_i)$ where $U_i$ is an open subset of M and $h_i$ is a homeomorphism of $U_i$

onto an open subset of $R^m$ such that (i) $\cup U_i = M$ and (ii) for every pair (i,j) the map $h_i \circ h_j^{-1} : h_j(U_i \cap U_j) \to R^m$ is a differentiable (of class $C^\infty$) map. The elements of the atlas are called local coordinate systems (charts) on M. A differentiable atlas is said to be maximal if we cannot add more charts to it still preserving the compatibility condition (ii). From any given differentiable atlas we can get a unique maximal differentiable atlas. Two differentiable atlases are equivalent if they give rise to the same maximal atlas.

DEFINITION I.1.2.

A differentiable ($C^\infty$) manifold of dimension m is a topological m-manifold M together with an equivalence class of maximal differentiable atlases.

Differentiable manifolds are also referred to as smooth or $C^\infty$ manifolds. Let $f: M \to N$ be a continous map between differentiable manifolds. Then f is said to be differentiable if for every i,j the map $g_j \circ f \circ h_i^{-1}$ is of class $C^\infty$ where $(U_i, h_i)$ is a coordinate system on M and $(V_j, g_j)$ is a coordinate system on N such that $h_i(U_i) \subseteq V_j$. Differentiable manifolds and Differentiable maps form a category. A differentiable map with a differentiable inverse is called a diffeomorphism. Two manifolds are diffeomorphic if there exists a diffeomorphism between them.

In the definition of an atlas, by requiring that the maps in condition (ii) are only differentiable maps of class $C^k$, $0 \leq k < \infty$, we get the notion of a $C^k$ atlas on M, and a $C^k$ manifold is analogously defined. A $C^o$ manifold is just a topological manifold. If the maps in (ii) are real analytic we get the notion of a real analytic ($C^\omega$) manifold. It is clear that a $C^r$-manifold is a $C^s$ manifold for $s \leq r$. Conversely, given a $C^r$-manifold it is possible to increase its differentiability. In fact any $C^r$-manifold, $r \geq 1$ admits a unique compatible real analytic manifold structure. This follows from Morrey [624] and Grauert [292]. But the existence of a $C^1$-manifold structure itself cannot be taken for granted. In 1960 Kervaire [456] gave the first example of a topological manifold (actually a PL manifold) of dimension 10 which does not admit any $C^1$-manifold structure.

Lie groups The situation in the case of Lie groups is pleasantly different and it is worth mentioning here.

DEFINITION I.1.3.

A Lie group G is a topological group G such that there exists a real analytic structure on the underlying set compatible with its topology so that G becomes a real analytic manifold and the maps of $G + G \to G$, $G \to G$ given by $(x,y) \mapsto xy$ and $x \mapsto x^{-1}$ are real analytic.

In the above definition, if we replace real analytic maps by $C^k$ maps we get the notion of a $C^k$-group, $0 \leq k \leq \infty$. As in the case of differentiable manifolds, one can ask whether on a $C^k$-group G there exists a compatible real analytic structure so that it becomes a Lie group. Hilbert asked for more!! Whether every $C^0$-group G admits a compatible Lie group structure? This is Hilbert's Fifth Problem. As a result of many years of investigation, the answer is now known in the affirmative. For a history of the solution of this problem see the recent survey of Yang [984] and for proofs in some important particular special cases see Montgomery and Zippin [618]. The uniqueness of the solution was known much earlier (for a proof see e.g. Varadarajan [934], chapter I, Section 2.6). Another pleasant fact of Lie groups is that any continuous homomorphism of Lie groups is actually real analytic. (For proof, see any book on Lie groups). But it should be noted that the existence of Lie group structures is far more unlikely than the existence of differentiable structures. For example while all the n-spheres $S^n$ are differentiable (real analytic) manifolds, only $S^1$ and $S^3$ are Lie groups.

Lie algebra of vector fields: Let M be a $C^\infty$ manifold. A vector field X is a section of the tangent bundle $TM \to M$. If U is a subset of M, X associates to each point $x \in U$, a tangent vector $X_x$ of M at x. Let f be a $C^\infty$ function on M. We define a function Xf by $Xf(x) = X_x(f)$. X is said to be $C^\infty$ (smooth) if Xf is $C^\infty$ for every $C^\infty$ function f. Let H(M) be the set of all $C^\infty$ vector fields on M. There is defined a bracket operator [ , ] in H(M); $[X,Y]f = X(Yf) - Y(Xf)$ for every $C^\infty$ function f on M. H(M) is an infinite dimensional Lie algebra over $\mathbb{R}$. If $\phi: M \to N$ is a differentiable map of differential manifolds, then it induces a map $\phi_*: T(M) \to T(N)$ by $(\phi_* X)_x = d\,\varphi_y(X_y)$ when $y = \phi(x)$ and d is the exterior derivative. $\phi_*$ is an algebra homeomorphism. If $\phi$ is a diffeomorphism of M onto N, then $\phi_*$ is a Lie algebra isomorphism of H(M) onto H(N). Thus diffeomorphic manifolds have isomorphic Lie algebras of vector fields. The Lie algebra structure of H(M) determines the differentiable structure of M, if M is compact (for proof see Shanks-Pursell [845], Amemiya-Masuda, Shiga [23], Masuda [575]). A fundamental question is whether it is possible to 'reconstruct' the differentiable structure of M from the Lie algebra structure H(M). Recently Grabowski [291] has shown that it is possible to 'reconstruct' the underlying set of M from H(M).

## I-A-1-(3): Piecewise Linear Manifolds (PL Manifolds).

Piecewise linear manifolds (briefly referred to as PL manifolds) form an important class between topological and differentiable manifolds. We assume in this section basic knowledge of piecewise linear topology. Good references for PL topology are Hudson [416] and Rourke-Sanderson [797].

A PL atlas on a topological m manifold is a family $(U_i, h_i)$, as in the definitions of a differentiable atlas, with the requirement that the maps $h_i$ be piecewise linear and $h_i \circ h_j^{-1}$ be PL isomorphic for each i, j such that $U_i \cap U_j \neq \phi$. Analogous to the definition of a differentiable manifold we formulate the definitions of a PL manifold and PL map between two PL manifolds. PL manifolds and PL maps form a category. Two PL manifolds are PL isomorphic if there exists a PL homeomorphism between them. (Note in the differentiable case a differentiable map which is a homeomorphism need not be a diffeomorphism).

There is a closely related notion of a combinatorial manifold. Let $\sigma_1, \sigma_2$ be two simplices in $\mathbb{R}^m$. We say that $\sigma_1$ and $\sigma_2$ are joinable if the set of all their vertices is independent. If this is the case, this set defines a simplex called the join of $\sigma_1$ and $\sigma_2$ and is denoted by $\sigma_1 . \sigma_2$. Let $K_1, K_2$ be two simplicial complexes. Then they are joinable if any simplex of $K_1$ is joinable to any simplex of $K_2$ and if $(\sigma_1, \sigma_2)$, $(\sigma_1', \sigma_2')$ are two joinable pairs $\sigma_1, \sigma_1' \in K_1, \sigma_2, \sigma_2' \in K_2$, then either $\sigma_1 \sigma_2 \cap \sigma_1' \sigma_2' = \phi$ or it is a face of both $\sigma_1 \sigma_1'$ and $\sigma_2 \sigma_2'$. If $K_1, K_2$ are joinable, then their join is the simplical complex $K_1 K_2 = K_1 \cup K_2 \cup \{\sigma_1 . \sigma_2, \sigma_1 \in K_1, \sigma_2 \in K_2\}$. Let $\sigma_1 \in K_1$ then its link is defined to be the complex, link $(\sigma_1, K_1) = \{\sigma_1' \in K_1 \mid \sigma_1'$ and $\sigma_1$ are joinable and $\sigma_1' . \sigma_1 \in K_1\}$. As usual for any complex K in $\mathbb{R}^m$, by $|K|$ we mean the polyhedron of K which is the union of all elements of K with the subspace topology induced from $R^m$.

## DEFINITION I.1.4.

A simplicial complex K is a combinatorial manifold of dimension m, if for all simplex $\sigma \in K$ the polyhedron $|\text{link}(\sigma, K)|$ is PL isomorphic to a PL sphere or a

PL ball of dimension $(m - \dim \sigma - 1)$.

The relation between the notions of PL and combinatorial manifolds is that if K is a combinatorial manifold of dimension m, then $|K|$ is a PL manifold of dimension m and conversely (for proof see Hudson [416]). In this sense we regard PL manifolds and combinatorial manifolds as equivalent notions.

DEFINITION I.1.5.

A topological manifold M is said to be triangulable if there exists a homeomorphism $f: |K| \to M$ where K is a simplicial complex. A triangulable manifold with a chosen triangulation is called a triangulated manifold. A differentiable triangulation of a differentiable manifold is a triangulation f such that f, restricted to each simplex, is differentiable. Similarly a PL triangulation is defined.

I-1-A-(4) Relations between the structures

It is known that every smooth manifold can be triangulated (Cairns). More precisely the existence and uniqueness of a compatible differentiable triangulation is known (Whitehead [970]). Consequently underlying a smooth manifold there exists a unique PL manifold. It follows immediately from the definition that underlying a PL manifold there exists a unique triangulated manifold and underlying a triangulated manifold there exists a unique topological manifold.

Do the reverse implications hold? It is well known that every topological 2-manifold can be triangulated (Rado) and that every topological 3-manifold can be triangulated, (Moise) (for proofs of both the statements see Moise [608]). It is not known whether every higher dimensional topological manifold admits a triangulation. It is also not known whether every triangulated manifold should admit a PL structure. The important discovery of Kervaire that we mentioned earlier is that there exists a PL manifold of dimension 10, such that it does not carry any compatible differentiable structure. Subsequently Eells and Kuiper [222] gave an example of an 8-dimensional PL manifold whose homotopy type does not contain any differentiable structure. It is known that any contractible PL manifold can have at most one compatible $C^\infty$ structure (Munkres [649], Thom [899]) and Hirsh [372] proved that every contractible PL manifold does indeed admit a compatible differentiable structure.

A great discovery of Milnor [594] is that in dimension $\geq 6$, there exist two homeomorphic polyhedra which are combinationally distinct thereby disproving the famous Hauptvermutung. In view of its basic importance, we give a brief account of this hypothesis.

The Hauptvermutung (The Fundamental hypothesis):

Let X be a topological space and $f : |K| \to X$ be a triangulation. Poincare associated certain integers $b_i(K)$ using the combinatorial properties of K. These were later proved by Alexander [20], to be independent of the triangulation but dependent only on X. These integers are denoted by $b_i(X)$ and are called the Betti numbers of X. It is pertinent to recall the well known formula $\Sigma(-1)^i b_i(X) = \Sigma(-1)^i r_i$ where $r_i$ are the number of i-dimensional simplices in a triangulation of X. From this the topological invariance of the homology groups can be proved.

Let $K_1, K_2$ be two simplicial complexes. The Hauptvermutung is that the existence of a homeomorphism between $|K_1|$ and $|K_2|$ implies the existence of a piecewise linear homeomorphism between them. If this hypothesis were true the topological invariance of homology would have followed. Alexander's proof of topological invariance of homology groups was different. But this hypothesis has been proved correct only in some special cases. It is true for polyhedra of dimension 3. It is true for manifolds of dimension 2 and 3. It is true if $K_1$ and $K_2$ are smooth triangulations of diffeomorphic manifolds. The important discovery of Milnor we mentioned earlier is that there exists, in each dimension greater than or equal to six, two homeomorphic but combinatorially distinct finite simplicial complexes $K_1, K_2$ thereby disproving the Hauptvermutung. In fact he proved the hypothesis in a strong sense in that no finite cell division of $K_1$ is isomorphic to a cell subdivision of $K_2$ (which implies the nonexistence of any piecewise linear isomorphism). Kirby-Siebenmann disproved the Hauptvermutung for higher dimensional topological manifolds. Smale proved the hypothesis to be true for combinatorial manifolds of dimension at least six. The best reference for details on this hypothesis and related questions is the book by Kirby-Siebenmann [463].

Exotic Spheres

An earlier discovery of Milnor [591] is of greater importance. In 1956 he discovered that on the 7-sphere $S^7$ (with its standard differentiable structure as a closed subspace of $R^8$) there exist at least seven differentiable structures all of which are homeomorphic to $S^7$ but no two of which are diffeomorphic. The existence of such structures was so unexpected they were called 'exotic spheres'. An exotic n-sphere is defined to be a compact connected n dimensional differentiable manifold which is homeomorphic to $S^n$ but not diffeomorphic to $S^n$. Kervaire-Milnor in their beautiful paper [457] proved that the number of diffeomorphism classes of differentiable structures on $S^n$, $n \neq 3$, is finite and these authors actually computed this number. The number of diffeomorphism classes of differentiable structures on $S^7$ is 28. We return to the work of Kervaire-Milnor again and give an interpretation in terms of Thom's cobordism theorem.

### I-1-A-(5) Poincare's conjecture and the topological classification Problem

We know that $S^1$ is the only compact connected topological 1 - manifold. Two connected compact topological 2-manifolds are homeomorphic if and only if they are both orientable or both non-orientable and have the same Euler characteristic. (For proof see e.g. Massey [572]). It follows that any simply connected compact topological 2-manifold is homeomorphic to the 2-sphere $S^2$. This led Poincare to conjecture:

#### Poincare's conjecture

Any simply connected compact topological 3-manifold is homeomorphic to the 3-sphere $S^3$.

This is a famous and important long-standing conjecture. In spite of the concentrated efforts of many great mathematicians for a long time, this conjecture remains unsettled. It is therefore natural to consider weaker forms of this conjecture by restricting the hypothesis and trying to solve them. Recall that a homotopy n-sphere is a compact topological n-manifold which has the same homotopy type as $S^n$. Poincare's conjecture is that any homotopy 3-sphere is homeomorphic to $S^3$. Since it has been found that topological manifolds are difficult to deal with it is better to restrict the study to a differentiable PL homotopy sphere. In view of the existence of the exotic structures, we cannot expect a differentiable homotopy n-sphere to be diffeomorphic to $S^n$ but can only hope that it is homeomorphic to $S^n$. This consideration leads to the:

Generalised Poincare conjecture

A differentiable homotopy n-sphere is homeomorphic to $S^n$.

A great achievement of recent times is the solution of this conjecture in the affirmative for $n \geq 5$. Smale, Zeeman and Stallings proved independently in quick succession that this conjecture is true for both PL and differentiable manifolds where $n \geq 5$. Later the conjecture has been proved by Newman to be true in the topological case also. The conjecture remains unsettled for $n = 3$ and 4 even in the differentiable case.

The Topological Classification Problem

The problem is to give a complete list of compact topological n-manifolds, $M_1$, $M_2$, ..., such that every homeomorphism class is represented once and only once in this list and any given compact topological n-manifold is homeomorphic to a manifold in the list. It is also important to give an algorithm to decide when two compact topological n-manifolds are homeomorphic (this is called the Homeomorphism Problem). As we have already mentioned, these two problems are solved in dimensions 1 and 2. The stumbling block in dimension 3 is the Poincare conjecture. Markov proved that the Homeomorphism Problem is unsolvable in dimensions $\geq 4$. It should be clear, therefore, why there has been intense study of 3-manifolds in recent years. The readers are referred particularly to Hempel [360] and Stallings [868]. The author has learned that Thurston has developed a theory to classify 'sufficiently large' 3-manifolds.

I-1-A(6) Obstruction Theories

What are the obstructions to the existence of triangulation of a topological manifold? We would like to mention an important result of Galewski-Stern [264].

THEOREM 1 (The Obstruction Theorem For Triangulations)

Every topological m-manifold $m \geq 6$, can be triangulated if and only if there exists a homology 3-sphere $H^3$ (that is a differentiable compact 3-manifold whose integral homology is that of $S^3$) with the following properties.

(1) $H^3 \# H^3$ bounds a PL 4-manifold ( # preserving orientation)

(2) Roh $(H^3) = 1$ where Roh denotes the Rohlin invariant of $H^3$.

(3) $H^3$ has the suspension property.

In the above # denotes the connected sum operation which is defined in the next subsection. For the notion of a Rohlin invariant see Rohlin [792], [793]. Recently Edwards [219] has proved that every homology 3-sphere has the suspension property. There will therefore be no obstruction for triangulating topological m-manifolds, $m \geq 6$, if there exists a homology 3-sphere with properties (1) and (2). Hence there has been a search going on for homology 3-spheres with properties (1) and (2). The readers are referred to the remarks of Siebenmann on page 64 in Browder [126].

What are the obstructions to the existence of a PL-structure on a topological manifold? We have the following important theorem of Kirby-Siebenmann [463] giving the single obstruction and classifying the PL structures.

THEOREM (The Obstruction And Classification Theorem For PL Manifolds)

Let M be a topological m-manifold, $m \geq 5$.

(1) There is one and only one well defined obstruction in $H^4(M; Z_2)$ to give a PL structure on M.

(2) Given a PL structure on M, the isotopy classes of PL structures on M are in 1-1 correspondence with the elements of $H^3(M; Z_2)$.

Using the above theorem Kirby-Siebenmann have shown that there are topological manifolds not admitting PL structures and there exist topological manifolds with distinct PL structures.

Given a PL manifold M can we construct a differentiable structure M' on M such that the underlying PL structure of M' is the given M? If this is possible we say that M is smoothable and M' is a compatible smooth structure. The study of the obstructions for smoothing a PL manifold and classifying (up to concordance) the compatible smooth structures has seen remarkable progress in recent years. Apart from Kirby-Siebenmann [463], some good references are Cairns [132], Hirsh [372], Hirsh and Mazur [373], Munkres [648], [649], [650], Lashof-Rothenberg [524], [525].

There is a subtler problem studied by Morita [621]. Let $M_{top}$ and $M_{PL}$ be respectively the underlying topological and PL manifolds of a given differentiable manifold M. Assume there is another PL structure $M_{top,PL}$ on $M_{top}$. Is it possible to smooth $M_{top, PL}$ compatibly? Morita has studied the obstructions and obtained some interesting results.

There is also an equivarient smoothing problem: given a finite group G acting on a topological manifold M the problem is to introduce a differentiable structure on M such that G acts differentiably? See Lashof [523] for details of this problem.

I-1-A-(7) Classification of differentiable manifolds in dimension $\geq 5$: Thom's Cobordism Theorem And Smale's h-cobordism Theorem.

In his fundamental paper [898], Thom developed a classification theory for compact differentiable manifolds. He introduced the important concept of cobordism which is weaker than the concept of diffeomorphism, and classified differentiable manifolds up to cobordism. The theory of Intrinsic Homology of Pontrjagin and Rohlin is a forerunner to the Cobordism theory of Thom.

DEFINITION I.1.7.1.

Let $M_1, M_2$ be two compact connected differentiable m-manifolds (without boundary). They are said to be cobordant if there exists a differentiable (m + 1)-manifold P with boundary $\partial P$ such that $\partial P = M_1 \cup M_2$ (disjoint union).

It is easy to check that cobordism is an equivalence relation and that two diffeomorphic manifolds are cobordant. We denote a cobordism between $M_1$ and $M_2$ by a triple $(P; M_1, M_2)$. Let [M] denote the equivalence class consisting of compact connected differentiable m-manifolds cobordant to a given compact connected differentiable m-manifold M. Let $\Sigma_m$ denote the set of equivalent classes of compact connected differentiable m-manifolds under cobordism.

It is easy to check that $\Sigma_m$ is an abelian group under the natural addition: $[M^m + N^m] = [M^m \cup N^m]$; its zero element is $[S^m]$ and every element of $\Sigma_m$ has order 2. Let $\Sigma = \bigoplus_{n \geq 0} \Sigma_m$. By defining a product in $\Sigma$ by $[M^m] \times [N^n] = [M^m \times N^n]$, $\Sigma$ gets a graded ring structure. Thom showed that $\Sigma$ is a graded polynomial algebra over $Z_2$ with one generator in each positive dimension m, except for those m of the form $2^n - 1$. He showed that for m even the real projective space $P^m(R)$ is a generator for this algebra. Later Dold [198] constructed odd dimensional generators. Using the structure of $\Sigma$, Thom proved the following celebrated theorem.

THEOREM 3 (Thom's Cobordism Theorem)

Two compact connected differentiable manifolds are cobordant if and only if their corresponding Stiefel-Whitney numbers are equal.

Stiefel-Whitney numbers of a manifold are computable, though not easily. Hence the above theorem is a good classification theorem.

DEFINITION I.1.7.1.

Let $M_1, M_2$ be two cobordant m-manifolds. Let $(P; M_1, M_2)$ be a cobordism. Then $M_1, M_2$ are said to be h-cobordant if the natural inclusions $M_1 \hookrightarrow P$, $M_2 \hookrightarrow P$ are homotopy equivalences.

It is easy to check that h-cobordism is an equivalence relation. We now state the h-cobordism theorem of Smale [861] which is another landmark in differential topology.

THEOREM 4 (Smale's h-cobordism Theorem)

Let $M_1, M_2$ be two h-cobordant n-manifolds. Let $(P; M_1, M_2)$ be an h-cobordism. If $n \geq 5$ and if $M_1, M_2$ and P are simply connected, then P is diffeomorphic to $M_1 \times [0,1]$ (or $M_2 \times [0,1]$).

Corollary 1 Two h-cobordant simply connected n-manifolds, $n \geq 5$, are diffeomorphic. (This is immediate from the theorem).

Remark h-cobordant manifolds are not homeomorphic in general. See Farrell and Hsiang [238].

Corollary 2 The generalised Poincare conjecture is true for $n \geq 5$. (The proof of the corollary is difficult).

There are many other important consequences of this theorem. In the light of this theorem we can give an interpretation of the main result of Kervaire-Milnor [451]. First we observe that h-cobordism classes form a group under 'the connected sum' operation, which we explain now.

Let X, Y be two oriented differentiable manifolds each of dimension n. Let $D^n$ be the n-disc. Let $f_1 : D^n \to X$ be an orientation preserving embedding and $f_2 : D^n \to Y$ be an orientation reversing embedding. For all $x \in \partial D^n = S^{n-1}$ identify $f_1(x)$ with $f_2(x)$. Then the disjoint union $x - f_1(\dot{D}^n) \cup Y - f_2(\dot{D}^n)$, where $\dot{D}^n$

denotes the interior of $D^n$, is called the connected sum of X and Y and is denoted by X # Y. It is independent, up to diffeomorphism, of the embeddings chosen to define it. It is an oriented differentiable manifold of dimension n. Let $\Gamma_n$ denote the set of all oriented h-cobordism classes of oriented homotopy n-spheres. This is a group with identity $[S^n]$; the inverse of any element is got by reversing the orientation. If $n \geq 5$, each homotopy n-sphere is homeomorphic to $S^n$ and two h-cobordant manifolds are actually diffeomorphic (Corollaries of Smale's h-cobordism theorem). Hence the order of the group $\Gamma_n$ gives the number of distinct differentiable structures that exist on $S^n$. This was computed by Kervaire-Milnor [457]. We have listed below the values of $|\Gamma_n|$ up to n = 16. The value $|\Gamma_3|$ is not known.

| n | 1 | 2 | 3 | 4 | 5 | 6 | 7 | 8 | 9 | 10 | 11 | 12 | 13 | 14 | 15 | 16 |
|---|---|---|---|---|---|---|---|---|---|---|---|---|---|---|---|---|
| $\lvert\Gamma_n\rvert$ | 1 | 1 | ? | 1 | 1 | 1 | 28 | 2 | 8 | 6 | 992 | 1 | 3 | 2 | 16256 | 2 |

I-1-A-(8) Classification of differentiable 4-manifolds

"It is only in the dimension of space-time, dimension 4, that the work of the century has badly failed to give an overview of the classification" remarks Siebenmann (page 63, Browder [126]). The main reason for this is that in dimension 4 'the techniques of geometric topology almost break down' as observed by Wall (page 65, Browder [126]). However there are some important results available on differentiable 4-manifolds and these results were obtained by studying quadratic forms associated with compact oriented differentiable 4-manifolds. An excellent introduction to this topic is Hirzebruch-Neumann-Kohn [387].

Let M be a compact connected oriented differentiable manifold of dimension 4k. Consider over $H^{2k}(M, \mathbb{R})$ the quadratic form $Q(\alpha,\beta) = (\alpha \cup \beta)[M]$ where $\alpha, \beta \in H^{2k}(M,R)$ and $\alpha \cup \beta$ denotes their cup product. Since Q is a bilinear symmetric form over a real vector space, it can be diagonalised. Consider a diagonalisation of Q and let $p^+$ be the number of positive entries and $p^-$ the number of negative entries. The index of Q, denoted by $\sigma(Q)$, is $(p^+ - p^-)$. Then index of M (also called the signature of M) is defined to be the index of the quadratic form associated with it and is denoted by $\sigma(M)$. The whole idea here is to study the geometry of M by studying the associated quadratic form; in particular the index $\sigma(M)$ gives a lot of information on M. The fundamental theorem, called the signature theorem (or the index theorem) is that $\sigma(M)$ can be expressed by a 'universal linear combination' of Pontrjagin numbers of M (see Hirzebruch [379] for the precise statement and proof). This theorem, its variants and the index theorems of Atiyah-Singer et alia have had a tremendous influence in modern mathematics; it is outside the scope and intentions of these notes to report on this matter. We refer the readers to the delightful lecture of Hirzebruch [382] to get an idea of these advancements.

An important result on the index of a compact connected oriented differentiable 4-manifold is

THEOREM 5 (Rohlin)

Let M be a compact connected oriented differentiable 4-manifold without a boundary. Let further the second Stiefel-Whitney number of M be zero. Then $\sigma(M) = 0 \bmod 16$.

For a detailed proof of this theorem and its relevance to the work of Kervaire-Milnor see Hirzebruch-Neumann-Kohn [387].

Much earlier than this work, Pontrjagin [743] and Whitehead [971] studied 4-manifolds in terms of the associated quadratic forms.

THEOREM 6 (Pontrjagin-Whitehead)

Let $M_1, M_2$ be two compact simply connected 4-manifolds. Let $Q_i$, $i = 1,2$ be the associated symmetric bilinear form on $H^2(M_i;Z)$ induced by the cup product $H^2(M_i,Z) \times H^2(M_i,Z) \to Z$. Then $M_1, M_2$ are homotopically equivalent if and only if $Q_1$ is congruent to $Q_2$.

For the notion of a congruent symmetric bilinear form see, for instance, Serre [835]. Hence the classification of 4-manifolds up to homotopy reduces to the classification of symmetric unimodular integral matrices which is known. See Milnor-Husemoller [601] and Serre [835]. Novikov [703] and Wall [945] strengthened the above theorem: two such manifolds $M_1, M_2$ are in fact h-cobordant. Since the h-cobordism theorem applies only when $\geq 5$, we cannot conclude that $M_1$ and $M_2$ are diffeomorphic. But using Smale's work on the generalised Poincare conjecture Wall [945] has proved the following theorem.

THEOREM 7 (Wall)

Let $M_1, M_2$ be simply connected compact differentiable 4-manifolds which are h-cobordant. Then there exists an integer $k > 0$, such that $M_1 \# k(S^2 \times S^2)$ is diffeomorphic to $M_2 \# k(S^2 \times S^2)$ (Notation: for any manifold X, kX denotes $X \# X \# \ldots \# X$ (k times)).

One natural question to ask in this connection is whether there exists any bound on the integer k. Recently Moishezen has given some estimates on k under the assumption that these manifolds admit complex structures. Also see Mandelbaum-Moishezen [566] [567]. Regarding the classification problem of 4-manifolds, it is pertinent here to refer the readers to the remarks in pages 63 to 66 in Browder [126].

The major results that we have described in this subsection are important, not only because they are landmarks as such, but also because of the fact that the essentially new concepts and difficult techniques that were employed in formulating and proving them enabled further advancements to be made. For example, the technique of surgery introduced by Kervaire-Milnor has turned out to be a powerful tool ; the concept of the microbundle of a topological manifold (analogue of the tangent bundle of a differentiable manifold) introduced by Milnor helped to generalise certain important results from differentiable manifolds to topological manifolds, and Thom's introduction of rational Pontrjagin classes for PL manifolds generalising the integral Pontrjagin classes enabled the study of PL manifolds to be undertaken more thoroughly (Novikov [704] proved that these rational Pontrjagin classes are actually topological invariants) ; Smale's generalisation of Morse theory enabled a deeper study of the topology of differentiable manifolds. The introduction of topological K-theory by Atiyah and Hirzebruch has had a tremendous influence in different branches of modern mathematics.

## I-1-A- (9) Eilenberg Maclane Complex and Aspherical Manifolds

For a very special class of closed topological manifolds there is a homotopy-theoretic analogue, namely, the Eilenberg-Maclane complexes $K(\pi, n)$ which are the building blocks of Homotopy theory. Eilenberg Maclane complexes are rightly regarded as the primitive objects of study for Homotopy theory since any CW complex can be built from these (Cartan-Serre [144], Whitehead [969 ]). The homotopy type of a topological space X is determined by its $K(\pi_n(X), n)$ constituents and its Postnikov invariants. A good reference for the topological study of CW complexes is Lundell and Weingram [552]. G.W. Whitehead [969], brings out in addition the fundamental role of Eilenberg Maclane spaces in homotopy theory. The notion of a CW-complex is due to J.H.C. Whitehead. The spaces, now known as Eilenberg Maclane spaces, were introduced and studied by Eilenberg and Maclane in a series of papers [230], [231], [232], [233]. For a quick history of the evolution of this concept and its importance see section 4: <u>The Bar construction</u>: of the lively address of Saunders Maclane entitled 'Topology and Logic as a source of Algebra' (Bull. A.M.S. 82(1976), 1 - 40). Here we give a brief account of CW-complexes, Eilenberg Maclane spaces $K(\pi,n)$ and closed topological manifolds of homotopy type $K(\pi,1)$.

<u>CW complexes:</u> A Hausdorff topological space is said to be compactly generated if each subset Y of X such that $Y \cap K$ is closed for every compact subset K of X is itself closed. Compactly generated spaces and continuous maps form a category. Locally compact Hausdroff spaces and metrizable spaces are compactly generated. All closed subsets of a compactly generated space are compactly generated. It can be proved that a Hausdorff space is compactly generated if and only if it has the weak topology with respect to the collection of all its compact subsets. Let $\{X_n \mid n \geq 0\}$ be a sequence of Hausdorff topological spaces such that (i) for each n, $X_n$ is closed subspace of $X_{n+1}$ (ii) $X_n \cap X_m$ is a closed subset of $X_n$ for every n,m (iii) the topologies induced by $X_n$ and $X_m$ on $X_n \cap X_m$ are equivalent. Take $X = \bigcup_0^\infty X_n$ and prescribe the weak topology on X given by the subspaces $X_n$. X may not be Hausdorff. But if each $X_n$ is compactly generated X will be Hausdorff and moreover compactly generated. Assume that each $X_n$ is compactly generated and that for each n, $(X_n, X_{n+1})$ has the homotopy extension property (with respect to arbitrary topological spaces). We then say $\{X_n\}$ is a filtration of X.

Consider pairs (X,A) where X is a topological space and A is a compactly generated subspace of X. A relative CW decomposition of (X,A) is given by a filtration $\{X_n\}$ of X satisfying (a) $A \subset X_o$ (b) for $n \geq 0$, $X_n$ is a cellular extension of $X_{n-1}$. A pair (X,A) with a relative CW decomposition is called a relative CW complex. If (X,A) is a CW complex then X must be compactly generated.

$(X_n, X_{n-1})$ are called the n-cells of $(X,A)$ and the set $X_n$ is called the n-skeleton of $(X,A)$. $(X,A)$ is said to be finite if there are only a finite number of cells. Taking $A = \Phi$, we get the notion of a CW complex. For example $S^n$ is a CW-complex with one 0-cell and with one n-cell $S^n$. If $(X,A)$ is a relative CW complex, then $Y = X/A$ is a CW complex with $Y_n = X_n/A$ for $n \geq 0$.

If K is a simplicial complex and L a sub-complex of K, then $(|K|, |L|)$ is a relative CW complex whose n-skeleton is $|K_n \cup L|$. In analogy with simplicial complexes the notion of a sub-complex of a CW complex is defined, but in contrast it should be noted that closed cells of a CW complex and their boundaries are not necessarily sub-complexes.

Let $(X,A)$ be a relative CW complex. It is said to be n-connected if for every relative CW complex $(Y,B)$ with dimension $Y \leq n$, any map $F: (Y,B) \to (X,A)$ is homotopic to a map of Y into A. $(X,A)$ is said to be $\infty$-connected if it is n-connected for every positive integer n. If $(X,A)$ is $\infty$-connected, A must be a deformation retract of X. A map $f: (X,A) \to (Y,B)$ between relative CW complexes is cellular if $f(X_n) \subset Y_n$ for every n. CW complexes and cellular maps form a category. Any continuous map between relative CW complexes can be approximated by cellular maps (the cellular approximation theorem). Following Eilenberg-Steenrod we can define singular homology and cohomology theories for a CW complex. Given a relative CW complex $(X,A)$ and a map $f: A \to Y$ the basic problem is to extend the map to all of X. This has to be done step by step from a lower dimensional skeleton to the next higher dimensional skeleton; at each stage one meets with an obstruction.

Eilenberg analysed these obstructions and developed an obstruction theory to tackle the Extension problem. He gave necessary and sufficient conditions (1) for the solvability of the Extension problem (known in the literature of the Eilenberg Extension theorem). (2) for the solvability of the Homotopy Problem (when two maps are homotopic ?) (known in the literature as the Eilenberg Homotopy theorem.) and (3) for the one to one correspondence between the homotopy classes of extensions of a given map and the cohomology group $H^n(X,A; \pi_n(Y))$ of $(X,A)$ with coefficients in the (only) homotopy group of an Eilenberg Maclane complex of type $K(\pi,n)$. (Known in the literature as the Eilenberg Classification theorem.)

This brings us to the Eilenberg Maclane spaces.

Remark

Recently there has been a study of Equivariant CW complexes (equivariant under the action of a compact Lie group). S. Illaman [427], [428], has developed a theory of equivariant singular homology and cohomology groups. The study of CW complexes under group actions has been intensively studied. See Bredon [114], Hsiang [415] for general information on the topology of transformation groups.

Eilenberg Maclane Complexes $K(\pi,n)$

A classical theorem of Hurewicz says that if X, Y are two path connected spaces with the same $n^{th}$ homotopy group ($\pi_n(X) = \pi_n(Y)$) and all other homotopy groups are zero, then X and Y must be of the same homotopy type. Hence they have the same cohomology and homology groups in all dimensions. Suppose X and Y have the same fundamental group and all higher homotopy groups are zero. Then Eilenberg-Maclane showed that for each Abelian group A, the cohomologies and homologies are the corresponding group cohomologies and group homologies of $\pi_1(X)$ with coefficients in A:

$$H^n(X,A) \approx H^n(\pi_1(X), A)$$
$$H_n(X,A) \approx H_n(\pi_1(X), A).$$

Eilenberg and Maclane studied more generally topological spaces with one non-zero homotopy group. Let $\pi$ be a group and n a positive integer. A topological space X is said to be an Eilenberg Maclane space of type $K(\pi, n)$ if $\pi_n(X)$ is the only vanishing homotopy group of X and $\pi_n(X) = \pi$. An Eilenberg Maclane complex of the type $(\pi,n)$ is a CW complex which is an Eilenberg Maclane space of type $K(\pi,n)$. An Eilenberg Maclane complex of type $(\pi,n)$ is denoted by $K(\pi,n)$. We list now some very nice properties of $K(\pi,n)$.

1) Given a positive integer n, and a group $\pi$ (abelian if $n > 1$), there exists a $K(\pi,n)$ and two such $K(\pi,n)$'s are of the same homotopy type.

2) Let $\pi, \pi'$ be two abelian groups and n a positive integer and let $a: \pi \to \pi'$ be a group homorphism. Then there exists a unique homotopy class of maps $f: K(\pi,n) \to K(\pi',n)$ such that $\pi_n(f) = \alpha$.

3) The cohomology and homology groups of $K(\pi,n)$ depend only on $\pi$, n and the

coefficient group. (These are in principle computable, but in practice they may be difficult to compute in specific cases.) (See e.g. Maclane [558] for the difficulties involved.)

4) For any CW complex X, the set $[X, K(\pi,n)]$ of homotopy classes of maps from X to a $K(\pi,n)$ has the structure of an abelian group. Moreover $[X, K(\pi,n)] \approx H^n(X,\pi)$.

5) Any topological space X "can be built up" from the $K(\pi,n)$'s. The homotopy type of X is determined by these $K(\pi,n)$'s and by its Postnikov invariants (which are certain cohomology classes).

This is the sought after basic theorem; its proof is difficult.

For the precise definition of Postnikov invariants and the proof of the above results see Whitehead [969], Chapters V and IX.

6) Recently it has been shown that there exist classifying spaces for fibrations with $K(\pi,n)$ as fibres (See Baues [75]).

<u>$K(\pi,1)$ Manifolds</u> A closed topological m-manifold is called a $K(\pi,1)$ manifold if it has the same homotopy type as $K(\pi,1)$ (These are also called aspherical.) Some basic unsolved problems are the following:

<u>Problem 1</u> Are there two $K(\pi,1)$ manifolds which are not homeomorphic?

<u>Problem 2</u> Is $R^m$ the universal covering of $K(\pi,1)$ m-manifolds? (This is known to be true in some special cases; see Lee-Raymonds [530], Johnson [437].)

<u>Problem 3</u> Can we give a complete list of all the groups which could be the fundamental groups of $K(\pi,1)$-manifolds? (In the case of 3-manifolds the existence of a 'huge' list is attributed to W. Jaco.)

Recently Raymond-Wigner have constructed some examples of aspherical manifolds with fundamental groups being discrete. They have shown that a discrete group $\pi$ is the fundamental group of an aspherical manifold if and only if it has no torsion. This improves the earlier results of Conner-Raymond. For results on aspherical manifolds see Conner-Raymond [172], [173].

<u>Remark</u>

There has been considerable progress in characterising these finite groups which can be the fundamental groups of compact differentiable n-manifolds whose universal coverings are the standard spheres $S^n$ or the exotic spheres $\Sigma^n$. Such differentiable manifolds are referred to as smooth spherical space forms. See Thomas-Wall [909], Madsen-Thomas-Wall [560], I. Madsen [559] for details on these manifolds.

### I-1-A-(10) Poincare Complexes

We remarked at the beginning (Remark 5 following the definition of topological manifold) that it is very difficult to recognize topological manifolds among topological spaces. Hence there is a need for introducing and understanding structures more primitive than the structures of topological manifolds. It was found in early 1960 that a convenient primitive structure is that of a Poincare Complex. A Poincare complex is a space satisfying a strong form of the Poincare duality: it is an analogue of a closed oriented manifold in homotopy theory. The details of the theory of Poincare complexes can be found in Browder [123], [124]; Spivak [866] and Wall [949]. We follow Browder.

DEFINITION

An oriented Poincare Complex of dimension $m$ is a CW complex $X$ having (i) finitely generated Homology groups $H_n(X)$, in each dimension $n$ and (ii) a homology class $[X] \in H_m(X)$ such that $X \cap: H^q(X) \to H_{m-q}(X)$ is an isomorphism for all $q$. (Here $\cap$ denotes the cap product.)

We assume throughout that the CW complexes have the property (i). It is easy to see that any closed orientable topological manifold is a Poincare complex.

DEFINITION

An oriented Poincare pair is a CW pair $(X, Y)$ with a class $[X] \in H_m(X, Y)$ such that $[X] \cap: H^q(X) \approx H_{m-q}(X, Y)$ is an isomorphism for all $q$.

It is clear from the definition that if $(X, Y)$ is a Poincare pair, then $Y$ is a Poincare complex. Note that the definition of a Poincare pair is symmetric in the sense that $[X] \cap: H^q(X) \to H_{m-q}(X, Y)$ is an isomorphism for all $q$ if and only if $[X] \cap: H^q(X, Y) \to H_{m-q}(X)$ is an isomorphism for all $q$.

We remarked already about the concept of a microbundle introduced by Milnor for a topological manifold (which is the analogue of the tangent bundle of a differentiable manifold) and its importance in smoothing theories especially. The difficulty with a Poincare complex is that there is no suitable notion of a tangent bundle for a Poincare complex. However, Spivak in his thesis introduced the concept of 'the normal fibre space' of a Poincare complex which can be viewed as the analogue of the normal bundle of a differentiable manifold embedded in a Euclidean space. This notion has turned out to be a very convenient and appropriate concept in this context.

DEFINITION

The Spival normal fibre space of a Poincare m-complex X is a pair $(\xi,\alpha)$ where $\xi$ is a (k-1) spherical fibration

$$\begin{array}{ccc} \Sigma & \longrightarrow & E \\ & & \downarrow p \\ & & X \end{array}$$

with $\Sigma$ having the homotopy type of the sphere $S^{k-1}$, $k > m + 1$ and $\alpha \in \pi_{m+k}(T(\xi))$ such that $h(\alpha) \cap U = [X]$.

In the above definition $T(\xi)$ stands for the Thom complex of $\xi$ (which is, by definition, equal to $X \cup_p cE$), h is the Hurewicz map $h: \pi(T(\xi)) \to \pi(T(\xi))$, $U \in H^k(T(\xi))$ is the Thom class and $\cap$ denotes the cap product $\cap: H_{m+k}(T(\xi) \otimes H^k(T(\xi) \to H_m(X)$. Note the 'structure group' of the Spivak normal fibration is not a group; it is only a monoid $G_k$ formed of all homotopy equivalences of $S^{k-1}$ onto itself.

The following theorem gives the existence of Spivak normal fibration.

THEOREM 8 (Browder-Spivak-Wall)

For an oriented Poincare m-complex X and for $k > m+1$, there exists a Spivak normal k-fibration $(\xi,\alpha)$. If $(\eta,\beta)$ is another Spivak normal k-fibration, then there exists a unique homotopy equivalence (unique up to fibre homotopy) $b: \xi \to \eta$ such that $T(b)_*(\alpha) = \beta$ where $T(b)$ is the induced map of the Thom Complex.

As in the case of bundles over a manifold, there is a classifying space $BG_k$ for the Spivak normal k-fibrations over X. This classifying space is related to the usual classifying space as follows: Let $BO_k$ be the classifying space of the tangent bundle (whose structure group is the orthogonal group $O_k$) of a differentiable manifold M. By Milnor, there exist classifying spaces $BPL_k$, BTop for the tangent bundles of PL and topological manifolds. We have then the natural maps $BO_k \to BPL_k \to BTop_k \to BG_k$.

Given a k-vector bundle over a simplicial complex X, it is classified by a map $X \to BO_k$: the composite map $X \to BO_k \to BPL_k$ induces a PL bundle over X; by forgetting the PL-structure, we get the map $BPL_k \to BTop_k$ which gives a Top k-bundle over X. This Top k-bundle over X gives a Spivak normal fibration over X (by removing the zero section, the fibre $R^k$ is changed to $R^k-(o)$, which has the homotopy type of $S^{k-1}$). Now consider a differentiable manifold X and take any smooth triangulation of X. The tangent bundle of the resulting PL manifold is obtained by 'triangulating' the tangent bundle of the differentiable manifold. (Lashof-Rothenberg). The tangent bundle of a PL manifold remains unchanged when the PL manifold forgets its PL-structure. The difficulty is in the next stage: how to relate the tangent bundle of a topological manifold to the Spivak normal fibration of its underlying Poincare complex. This forces us 'to stabilise' the above implications by taking their direct limit as $k \to \infty$. Doing this, we get

$$BO \to BPL \to BTop \to BG.$$

Spivak has shown then that the tangent bundle of the topological manifold X induces a map $M \to BTop \to BG$ which is homotopic to that induced by 'an inverse' to the Spivak stable normal fibration of the Poincare complex X.

The fundamental problems are (A) Given a Poincare Complex X, is it possible to impose on X a compatible topological manifold structure? (PL-structure? or Differentiable structure?). That is: Is it possible to lift the Spivak normal fibration to a topological bundle (PL bundle or $B_O$ bundle)?

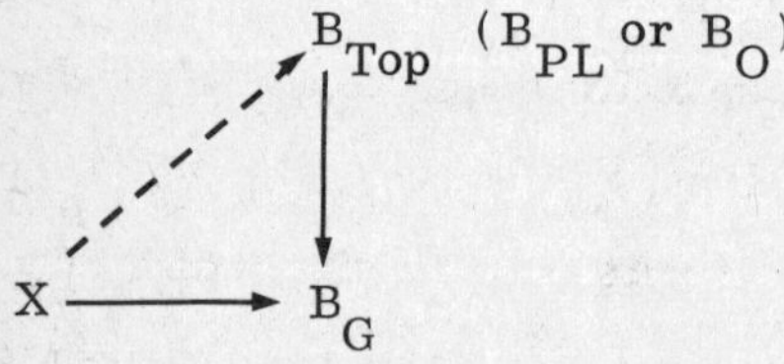


(B) If so, give a classification of such manifold structures (PL structures, Differentiable structures).

Regarding (B), the classification is sought for the concordance classes: Let $(X, \partial X)$ be a Poincare pair. Let $(M_1, h_1)$ be a pair where $M_1$ is a topological manifold and $h_1 : (M_1, \partial M_1) \to (X, \partial X)$ is a homotopy equivalence. Two such pairs $(M_1, h_1)$ and $(M_2, h_2)$ are called concordant if there exists a cobordism U, $\partial U =$

$M_1 \cup M_2 \cup V$, $\partial V = \partial M_1 \cup \partial M_2$ and a homotopy equivalence:

$$H : (U,V) \to (X \times [0,1], \partial X \times [0,1]) \text{ with}$$

$$H/M_i = h_i : M_i \to X \times (i), \; i = 0,1.$$

Analogously, concordance is defined in the PL and Differentiable cases.

Note that concordance is an h-cobordism. Hence if $\partial X = \pi_1(X) = 0$ and dim of $X \geq 5$, Smale's h-cobordism theorem tells us that $h_2^{-1} h_1$ is homotopic to a homeomorphism (PL isomorphism, diffeomorphism).

We have the following theorems regarding question (A).

THEOREM 9 (Kirby-Siebenmann)

Let X be an oriented Poincare m-complex ($m \geq 5$) and let $\pi_1(X) = 0$. Then X is of the homotopy type of a topological manifold if and only if the Spivak normal fibration is fibre homotopy equivalent to a Top-bundle.

In the PL case an analogous theorem holds (Browder-Hirsh), but in the differentiable case the situation is very complicated. We do not give here the precise statement of the theorem on the homotopy type of smooth manifolds, which is due to Novikov-Browder. It says that uder restrictive conditions on the Poincare m-complex $m \geq 5$, it is of the same homotopy type as a differentiable manifold. (For the precise statement of Novikov-Browder theorem see Browder [123], [124].

Let $(X, \partial X)$ be a Poincare pair (where $\partial X$ could be empty). Let $\Sigma_{Top}(X)$ ($\Sigma_{PL}(X)$, $\Sigma_O(X)$) be the set of all concordant classes of pairs $(M,h)$ where M is a topological manifold (PL-manifold, differentiable manifold) and $h : (M, \partial M) \to (X, \partial X)$ is a homotopy equivalence. Then we have the following desired classification theorem.

THEOREM 10 (Kirby-Siebenmann (Browder-Hirsh, Wall-Spivak-Sullivan))

Let $(X,Y)$ be an oriented Poincare pair of dimension $\geq 6$. Let $Y \neq \phi$ and X,Y be 1-connected. Then the elements of $\Sigma_{Top}(X)$ ($\Sigma_{PL}(X)$, $\Sigma_O(X)$) are in one to one correspondence with the homotopy classes of cross sections of $f^*(p_{Top})$ ($f^*(p_{PL})$, $f^*(p_O)$) where $f : X \to B_G$ ($f : X \to B_{PL}$, $f : X \to B_O$) is the classifying map of the Spivak normal fibration and $p_{Top} : B_{Top} \to B_G$ ($p_{PL} : B_{pL} \to B_G$, $p_O : B_O \to B_G$) is the natural fibration.

Given one topological structure ( PL-structure, differentiable structure ) on X, there exists an isomorphism

$\Sigma_{Top}(X) \approx [X, G/_{Top}]$ ( $\Sigma_{PL}(X) \approx [X,G/_{PL}]$, $\Sigma_{O}(X) \approx [X,G/_{O}]$ ).

Thus in this case the situation is completely analogous to the smoothing theory of PL manifolds. In the case $Y = \phi$, the situation is more difficult and is different from the smoothing theory of PL manifolds. In the non-simply connected cases the details involved in surgery (of non-simply connected manifolds) get very technical and complicated. The readers are referred to the book by Wall [950] for a classification theorem and for further details.

We summarize the entire discussion of this part by the following sequence of implications.

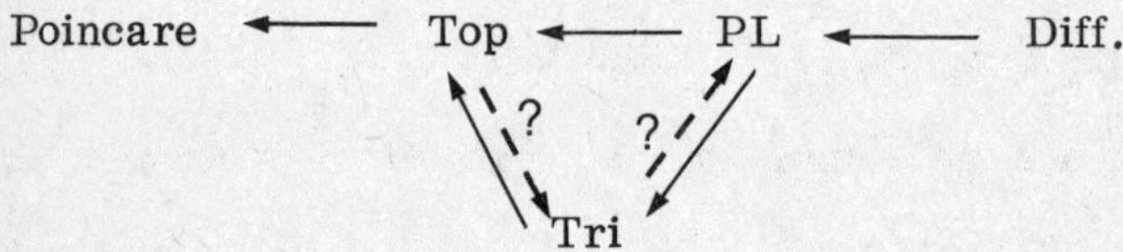


Section 1: Part B

## I-1-B-(1) Complex Manifolds

Let M be a para compact Hausdorff space. A holomorphic n-atlas on M is a family $(U_i, h_i)$ where $U_i$ are open subsets of M with $\cup U_i = M$, $h_i$ are homeomorphisms of $U_i$ onto open sets in $\mathbb{C}^n$ such that $h_i \circ h_j^{-1}$ are biholomorphic whenever $U_1 \cap U_j \neq \phi$. A maximal holomorphic n-atlas and equivalence of holomorphic n-atlases are defined as in the differentiable case. M together with an equivalence class of holomorphic n-atlases is called a complex manifold of dimension n.

It is clear that underlying a complex manifold M there exists a unique differentiable manifold $\underline{M}$. Let $T_x\underline{M}$ be the tangent space at x to $\underline{M}$, $T_x\underline{M} \otimes_R \mathbb{C}$ be its complexification and let $e_1, \ldots, e_n$ be a basis of this complex vector space. Then the vectors $e_1, \sqrt{-1}e_1, \ldots, e_n, \sqrt{-1}e_n$ can be chosen to be an oriented basis of the underlying real vector space: that is any two such bases differ from each other by a positive determinant. This orientation is independent of the choice of the basis and invariant under holomorphic automorphism since $GL(n, \mathbb{C})$ is connected. Hence the bundle $T\underline{M} \otimes_R \mathbb{C}$ acquires a canonical orientation. This gives rise to a canonical orientation in M. Thus any complex manifold has a canonical orientation.

Let M be a complex manifold with local coordinate system $h_i$. Let W be an open subset of M. A continuous map $f: W \to \mathbb{C}$ is said to be holomorphic if $f.h_i^{-1}$ is holomorphic for each i where defined. Let N be another complex manifold with local coordinate system $(\lambda_j)$. A map $f: M \to N$ is called holomorphic if $\lambda_j \circ f.h_i^{-1}$ is holomorphic, for each i and j, where defined. A holomorphic map is called biholomorphic if the inverse map $f^{-1}$ is also holomorphic. Complex manifolds and holomorphic maps form a category. Two complex manifolds are said to be holomorphically (or complex analytically) isomorphic if there exists a biholomorphic map between them. Thus we can talk of distinct (isomorphism classes of ) complex structures on a given differentiable manifold. But the existence of a complex structure itself is a nontrivial phenomenon as will be clear from the following discussion in this part.

As in the case of differentiable manifolds, the Lie algebra H(M) of all holomorphic vector fields of type (0,1) on a compact complex manifold M determines the complex structure of M completely, (see Amemiya [227]). A fundamental open problem is whether it is possible to 'reconstruct' the complex structure of M in terms of the Lie algebra structure of H (M). This question arises only if the corresponding construction is possible in the differentiable case.

### I-1-B-(2) Almost complex Manifolds

Let M be a differentiable manifold. An almost complex structure on M is a tensor field J (a differentiable section of the tangent bundle TM) such that at every $x \in M$, $J_x : T_x(M) \to T_x(M)$ is an endomorphism and $J_x^2 = -1$. (Such a $J_x$ is called a complex structure on the real vector space $T_x(M)$. An almost complex manifold is a differentiable manifold with an almost complex structure.

Let $M_1$ and $M_2$ be two almost complex manifolds with almost complex structures $J_1, J_2$ respectively. A mapping f is said to be an almost complex mapping if $J_2 f_* = f_* J_1$ where $f_*$ represents the induced mapping on the tensor bundles.

Almost complex manifolds and almost complex maps form a category. It is easy to check that every almost complex manifold must be of even dimension and orientable. Thus the obvious necessary conditions for an almost complex structure to exist on a differentiable manifold are that the manifold be of even dimension and be orientable. But these obvious necessary conditions are sufficient only in dimension 2. It is known that no even dimensional spheres other than $S^2$ and $S^6$ can admit any almost complex structure (as we shall indicate presently). An equivalent way to define an almost complex structure is to define it as a $GL(m, \mathbb{C})$ -structure (a precise definition is given later in this subsection). Then necessary conditions (not so obvious as the above conditions) for the existence of almost complex structures must be sought in terms of 'characteristic classes' (see Wu [981]). We have the following important theorem.

THEOREM 1

The necessary conditions for the existence of an almost complex structure on a differentiable manifold of dimension 2m are that

(1) the odd Stiefel-Whitney classes of the SO (2m) bundle be zero, and

(2) $$\sum_{O<i\leq [m/2]} (-1)^i p_i = \sum_{O\leq i\leq m} (-1)^i C_i \sum_{O<j\leq m} C_j ,$$

where $p_i$ are the Pontrjagin classes of the $SO(2m)$-bundle and $C_j$ are the Chern classes of the reduced $U(m)$-bundle (as usual $SO(2m)$ denotes the special orthogonal group of $2m \times 2m$ matrices, $U(m)$ denotes the unitary group of $m \times m$ matrices).

Apply the above theorem to $S^{4k}$. It is known that (1) is satisfied and all the $p_i$ are zero, hence the above formula gives $C_{2k} = 0$ if $S^{4k}$ admits an almost complex structure. But $C_{2k}$ can be proved to be the Euler class of $S^{4k}$ and hence cannot be zero. This proves that no $S^{4k}$ can admit any almost complex structure. We shall state an integrability theorem of Bott from which it will follow that no $S^{2n}$, except $n = 1,2,3$, can can admit any almost complex structure. Granting this we see that $S^2$ and $S^6$ are the only spheres which can admit almost complex structures. It is known that $S^2$ admits a unique almost complex structure. It is known that one can construct an almost complex structure on $S^6$ using Cayley numbers. (see for example Kobayashi-Nomizu, pages 139, 140, Volume 2, [475].)

Are there necessary and sufficient conditions for the existence of almost complex structures? In dimensions $\leq 10$, necessary and sufficient conditions are known (see Ehresmann [229], Hirzebruch-Hopf [385], Heaps [357]. We shall state in Section I-1-B-14 the necessary and sufficient conditions given by Hirzebruch and Hopf for the existence of an almost complex structure on 4-manifolds.

Here we state Heaps theorem, giving necessary and sufficient conditions for the existence of almost complex structures on 8-manifolds just to illustrate how complicated it could be in higher dimensions.

THEOREM 2 (Heaps)

Let M be a compact connected oriented differentiable manifold of dimension 8. Then M admits an almost complex structure if and only if

(a) $W_8(M) \in Sq^2 H^6(M; Z)$

and there exist cohomology classes $u \in H^2(M; Z)$ and $v \in H^6(M; Z)$ such that

(b) $\rho u = W_2(M)$, $\rho v = W_6(M)$

(c) $2\chi(M) + u.v = O \bmod 4$

(d) $8\chi(M) = 4p_2(M) + 8uv - u^4 + 2u^2 p_1(M) - p_1(M)^2$

Remarks

In the above $W_i$, $p_i$, $\chi$ denote respectively the Stiefel-Whitney, Pontrjagin and Euler class of M.

$Sq^2$ denotes the Steenrod square operation (see Steenrod [872]. $\rho$ denotes the mod 2 reduction.

Thomas [905] has shown that if $sq^2 H^6(M;Z) = O$ then (c) is a consequence of (a) and (b).

For some more interesting results on the obstructions to the existence of an almost complex structure see Massey [572], Morita [622]. Necessary and sufficient conditions in higher dimension ($> 10$) for the existence of almost complex structures seem not to be known.

I-1-B-(3) Integrable almost complex manifolds

Let M be a complex structure on a differentiable manifold M. Let $T_x(M)$ denote the complex tangent space of M at x and $T_x(\underline{M})$ be the real tangent space of M at x. Then it is easy to check that $T_x(\underline{M})$ is canonically isomorphic with the real vector space underlying the complex vector space $T_x(M)$. Hence the complex structure of $T_x(M)$ induces a complex structure $J_x$ on the real vector space $T_x(\underline{M})$

and we can check that $J_x$ varies smoothly (differentiably) with respect to x. Hence $J : TM \to TM$ is a vector bundle mapping such that $J^2 = -1$, where 1 denotes the identity bundle mapping. It can be checked that a map $f : M_1 \to M_2$ between complex manifolds is holomorphic if and only if f is an almost complex map with respect to the underlying almost complex structures of $M_1$ and $M_2$. Thus we see that underlying a complex manifold there exists a unique almost complex complex manifold. Conversely, given an almost complex manifold, does there exist a complex manifold whose underlying almost complex manifold is the given one? If so, we say that the given almost complex manifold is integrable. It is easy to verify that $S^2$ admits a unique almost complex structure and it is integrable.

It is known that there are almost complex manifolds which are not integrable and whose underlying differentiable manifolds cannot carry any integrable amost complex structures. Van de Ven [935] was the first to give such an example. We mentioned that $S^6$ admits an almost complex structure J constructed by means of Cayley numbers. Let us call this structure, for convenience, the Cayley structure on $S^6$. It is not difficult to check (see, for instance, Frolicher [253]) that the Cayley structure on $S^6$ is not integrable. It is not known whether $S^6$ can carry any other almost complex structure which is integrable. It is important here to refer to Adler [10] where it is claimed that $S^6$ does not admit any integrable almost complex structure on $S^6$. However the proof of this claim has been found to be in error. Hence whether $S^6$ admits a complex structure still remains a problem. The conjecture is it is not a complex manifold. (We take up the problem of integrability of almost complex structures in I-1-B-(12)). There are important notions of 'invariant almost complex structures' and integrable invariant almost complex structures in the study of homogenous spaces of Lie groups. We shall not discuss these here. We refer to Kozuel [501], Borel-Hirzebruch [103]. Kobayashi-Nomizu [475], Chapter X, Volume 2, and Helgason [359].

### I-1-B-(4) Van de Ven's example

In 1966, Van de Ven gave examples of Compact almost complex manifolds of dimension 4 which do not admit any integrable amost complex structures. Consider $M = \underline{IP^2(\mathbb{C})} \# S^1 \times S^3 \# S^1 \times S^3$ where $\underline{IP^2(\mathbb{C})}$ denotes the differentiable 4-manifold underlying the complex projective plane $IP^2(\mathbb{C})$ and # denotes the connected sum operation. As we shall see later, the differentiable manifold $S^1 \times S^3$ admits complex structures and we can actually characterize all the complex structures on $S^1 \times S^3$ (these are the Hopf surfaces). $\mathbb{P}^2(\mathbb{C})$ is a complex manifold. Thus M is obtained by taking the connected sum of three 4-manifolds each of which admits complex structures. However Van de Ven showed that M is an almost complex manifold but M can never be a complex manifold.

This is a very deep result. Van de Ven uses the Atiyah-Singer index theorem among other important results for proving this result. There are two things to be proved, that M admits an almost complex structure and that M does not admit any integrable almost complex structure. Using the sufficient condition of Hirzebruch Hopf for the existence of almost-complex structures on 4-manifolds (see section I-1-B-(14) it could be proved that M is almost complex. To see that M cannot admit any integrable almost complex structure assume, if possible, that M does admit integrable complex structure. Then M would be a compact connected complex surface and it would be even algebraic and by applying nontrivial results of algebraic surfaces we could conclude that the Euler characteristic of M is non-negative. But the Euler characteristic of M is -1. This contradiction proves that M cannot admit any integrable almost complex structures. For a detailed discussion on Van de Ven's examples see Pittie [733].

Van de Ven also mentions another example : Let

$$V = \{[Z_o, \dots Z_4] \in \mathbb{P}^3(\mathbb{C}) \mid Z_o^4 + \dots + Z_4^4 = O\}.$$

This is a simply connected compact differentiable 4-manifold. Consider $M = V \# \dots \# V$ ((2k + 1) times). Then M is an almost compex manifold but it is never a complex manifold. Recently Yau [988] has shown that $T^3(R) \# P^3(R) \times S^1$ does not admit any complex structure ($T^3(R)$ is a real 3-torus). The proofs of these assertions are also quite difficult.

The situation in the case of open manifolds (no component being compact) is pleasantly different, as was shown recently by Landweber [419].

<u>THEOREM 3 (Landweber)</u>

(i) Let X be any open differentiable manifold of dimension 2m such that $H^i(X, Z) = O$ for $i > m$, then every almost complex structure on X is homotopic to a complex structure.

(ii) Let X be as above with $H^i(X, Z) =$ for $i \geq m$. Then there is a natural one to one correspondence between the set of all almost complex structures on X and the set of all complex structures on X.

I-1-B-(5) Moishezon's conjecture

Let $\mathbb{P}^2(\mathbb{C})$ be the complex projective plane with its natural orientation and let $Q^2(\mathbb{C})$ denote $\mathbb{P}^2(\mathbb{C})$ with the orientation reversed. If X is any complex manifold there is a natural way of extending the complex structure of X to the connected sum $X \# Q^2(\mathbb{C})$ (by the blowing up operation which is explained in Chapter III, Section 2). But there is no natural way to extend the complex structure to $X \# \mathbb{P}^2(\mathbb{C})$. For any integer $\ell$, it is also possible to extend the complex structure to $X \# \ell Q^2(\mathbb{C})$ (Remember our notation that $\ell Q^2(\mathbb{C})$ means $Q^2(\mathbb{C}) \# \ldots \# Q^2(\mathbb{C})$, $\ell$ times). Now consider for any $k \geq 2$, any $\ell \geq 1$ the connected sum $Y = k\,\mathbb{P}^2(\mathbb{C}) \# \ell\, Q^2(\mathbb{C})$. Y is a 4-dimensional compact oriented differentiable manifold. It is known that there exist many pairs $(k, \ell)$ such that Y admits an almost complex structure.

Moishezon's conjecture

Y does not admit any complex structure.

For more details on this problem see Mandelbaum-Moishezon [566], [567].

I-1-B-(6) Hirzebruch Surfaces

From the foregoing discussion it should be clear that the existence of a complex structure is a highly nontrivial problem. Sometimes a differentiable manifold may admit a unique complex structure. For example $S^2$. (There is a conjecture that the complex structure on the complex projective space $\mathbb{P}^n(\mathbb{C})$ is unique. We take up this important question in the next section.) Hirzebruch [374] showed that there exist infinitely many distinct complex structures on $S^2 \times S^2$. Consider $\mathbb{P}^2(\mathbb{C}) \times \mathbb{P}^1(\mathbb{C})$. Let $(x_o, x_1, x_2)$ and $(y_1, y_2)$ be homogeneous coordinates in $\mathbb{P}^2(\mathbb{C})$ and $\mathbb{P}^1(\mathbb{C})$, respectively. Define surfaces $H_n$ in $\mathbb{P}^2(\mathbb{C}) \times \mathbb{P}^1(\mathbb{C})$ defined by the equation $x_1 y_1^n - x_2 y_2^n = 0$. Hirzebruch showed that the $H_n$ are all distinct as complex manifolds and for all even n they are all homeomorphic to $S^2 \times S^2$ (He also showed that for all odd n, they are homeomorphic but not to $S^2 \times S^2$. For details see, in addition to Hirzebruch [374]; pages 15 and 16 in Kodaira-Morrow [489]. It is not known whether $S^2 \times S^2$ admits complex structures other than the Hirzebruch structures. Andreotti [26] has shown that they do not admit any other algebraic structure.

I-1-B-(7) Special complex structures - Affine and Projective structures

A complex affine manifold is a complex manifold whose transition functions are affine. That is if $(z_i^1, \ldots, z_i^n)$ and $(z_j^1, \ldots, z_j^n)$ are local coordinates in $U_i, U_j$ respectively and if $U_i \cap U_j \neq \emptyset$, then letting $Z^{(i)}$ be the column vector $t(Z_i^1, \ldots Z_i^n)$ and $Z^{(j)}$ be the column vector $t(Z_j^1, \ldots, Z_j^n)$, we must have $Z^{(i)} = A_{ij} Z^{(j)} + b_{ij}$ where $A_{ij} \in GL(n, \mathbb{C})$ and $b_{ij} \in \mathbb{C}^n$. Not every compact complex manifold admits a complex affine structure. Among the one-dimensional compact connected complex manifolds (Riemann surfaces) only the tori admit complex affine structures. There exist countably many distinct complex affine structures on any complex one-dimensional torus (see Gunning [337]). Thus a fixed compact complex manifold may have many distinct complex affine structures on it. The complex affine structures can be characterized by the fact that there is a one-to-one correspondence between the complex affine structures on a complex manifold and its holomorphic linear connections with curvature and torsion both zero (see Matsushima [584], Vitter [941], Sakane [808].

More generally a complex projective structure is defined by requiring that the transition functions be projective transformations, (for a precise definition see Gunning [338], pages 50 and 55), belonging to a pseudo group characterized by a partial differential equation. Again not every compact complex manifold admits complex projective structures. But every Riemann surface of genus greater than one does admit projective structures and they have been classified (see Gunning [338]). Recently Gunning has taken up the problem of the existence and classification of projective structures on compact complex surfaces (complex manifolds of complex dimension 2) and has obtained important results. The readers are referred to the Lecture Notes of Gunning [338]. The problem of the existence and classification of complex projective structures in higher dimensions is much more difficult. For some interesting results in higher dimensions see Ganesh [266].

## I-1-B-(8) Exotic complex structures

Just as exotic differentiable structures exist so do 'exotic complex structures', that is there exist complex structures on a compact topological manifold whose underlying differentiable structures are exotic. Brieskorn and Van de Ven [120] gave such examples in 1968. They considered $S^1 \times \Sigma^{2n-1}$, where $\Sigma^{2n-1}$ is an exotic (2n-1) sphere ($\Sigma^{2n-1}$ is homeomorphic to $S^{2n-1}$ but not diffeomorphic to $S^{2n-1}$), and showed it carries complex structures provided $\Sigma^{2n-1}$ is a homotopy sphere forming the boundary of a parallelizable manifold, $n \neq 2$. (Recall that a differentiable m-manifold X is parallelizable provided its tangent bundle is trivial. $S^1, S^3, S^7$ are the only spheres which are parallelizable.) More generally they showed that $\Sigma^{2n-1} \times \Sigma^{2n-1}$ (each being homotopy sphere bounding a parallelizable manifold) carries complex structures.

We now explain how Brieskorn and Van de Ven constructed these interesting examples. Let $a = (a_o, \ldots a_n)$ be an (n+1) tuple of positive integers and let $(Z_o, \ldots Z_n) \in \mathbb{C}^n$. Let X(a) be the affine algebraic variety given by $\sum_{i=O}^{n} Z_i^{ai} = O$. It is easy to check that if some $a_i = 1$, then this variety X(a) is non-singular, otherwise the origin is the only singularity of X(a). Let $\Sigma(a) = S^{2n+1} \cap X(a)$. Then $\Sigma(a)$ is a differentiable (2n-1) manifold which is oriented (see, for example, Milnor [600]). If X(a) is non-singular, it is clear that $\Sigma(a)$ is $S^{2n-1}$. Brieskorn [117] has shown that all homotopy spheres $\Sigma^{2n-1}$, $n \neq 2$, bounding a parallelizable manifold appear among $\Sigma(a)$. In fact he showed there exist infinitely many a such that every (2n-1) dimensional homotopy sphere $\Sigma^{2n-1}$ bounding a parallelizable manifold is diffeomorphic to $\Sigma(a)$.

$\mathbb{C}$ acts holomorphically in the complex manifold X(a) - O by $t(z_o, \ldots, Z_n) \to (e^{t/a} Z_o, \ldots, e^{t/an} Z_n)$ and this defines the actions of $\mathbb{R}$ and Z on X(a) - O. Z acts properly discontinuously and freely on X(a)-O and hence, by a theorem of Cartan, X(a) - O/Z carries a natural complex manifold structure which we denote by H(a). H(1,...1) is called a Hopf manifold.

The map $(Z,t) \to t(Z)$ is a diffeomorphism of $\Sigma(a) \times \mathbb{R}$ onto $\Sigma(a) \times S^1$. Hence we conclude that there exist complex structures on $S^1 \times \Sigma^{2n-1}$, $n \neq 2$. Since exotic spheres appear among $\Sigma(a)$, we have the desired result showing that the underlying differentiable structures of these complex structures are exotic. By applying Smale's cobardism theorem which we stated in the previous section, it can be seen that if $\Sigma_1, \Sigma_2$ are homotopy spheres of dimension $\geq 5$, then $S^1 \times \Sigma_1$ is diffeomorphic to $S^1 \times \Sigma_2$ if and only if $\Sigma_1$ is diffeomorphic to $\Sigma_2$. The construction of complex structures on $\Sigma^{2n-1} \times \Sigma^{2n-1}$ is similar to the construction of complex structures on $S^1 \times \Sigma^{2n-1}$.

The topology of $\Sigma(a)$ has been studied in detail by Milnor; the difficulty is that $\Sigma(a)$ is embedded in 'a knotted manner' in $S^{2n+1}$ and these knots have to be analyzed very carefully. (See also Phan [731], Hirzebruch-Mayer [388].)

I-1-B-(9) Γ-Structures and G-Structures

The concepts of Γ-structures and G-structures are two general concepts which unify a number of significant geometric structures. Differentiable and complex manifolds become particular cases.

DEFINITION

Let D be an open set of $\mathbb{R}^n$. A pseudo group of transformations on D is a set Γ of local diffeomorphisms of D such that (i) $f \in \Gamma \Rightarrow f^{-1} \in \Gamma$ (ii) If $f \in \Gamma$, $g \in \Gamma$ are such that $g \circ f$ is defined then $g \circ f \in \Gamma$ (iii) Let f be any local diffeomorphism of D and let $D = \cup D_i$. Then $f \in \Gamma$ if and only if the restriction of f to each $D_i$ belongs to Γ (iv) The identity mapping of D belongs to Γ.

A pseudo group Γ is said to be transitive if for every pair of distinct points x, y in D, there exists an $f \in \Gamma$ such that $f(x) = y$. A pseudo group is a Lie pseudo group if the elements of Γ satisfy a partial differential equation (for the precise definition see Goldschmidt [286]).

The definition of a Γ-manifold is given analogously to the definition of a differentiable (or complex) manifold. Let X be a paracompact Hausdorff space and Γ be a pseudogroup of transformations in $\mathbb{R}^n$. A Γ-atlas on X is a collection $(U_i, h_i)$ of open sets $U_i$ of X such that $\cup U_i = X$ and $h_i$ are homeomorphisms of $U_i$ onto open sets of $\mathbb{R}^n$ such that $h_i \circ h_j^{-1} \in \Gamma$ for every i, j. A maximal Γ-atlas and the equivalence of Γ-atlases are defined as in the case of differentiable atlases. Then a Γ-manifold X is a paracompact Hausdorff space with an equivalence class of Γ-atlases. Let $\Gamma_d$ ($\Gamma_c$) be the set of all local diffeomorphisms of $\mathbb{R}^m$ (local biholomorphic maps of $\mathbb{C}^n$). Then a $\Gamma_d$-structure ($\Gamma_c$-structure) is a differentiable m-manifold (a complex n-manifold). By suitably choosing $\Gamma$, flat, affine and projective structures, symplectic structures, contact structures and f-structures (which are standard objects of study in differentiable geometry)can be defined as $\Gamma$-manifolds. We refer the readers to the book of Kobayashi [473] and the reference given there for detailed information on such structures.

The study of pseudo groups was stated by Eli Cartan [140], [141]. For modern accounts see Kuranishi [510], Spencer [865], Singer-Sternberg [857], Liebermann [545], Ehresmann [229].

## I-1-B-(10) G-Structure

Let X be any differentiable m-manifold. $TX \to X$ its tangent bundle and $Gl(x,R) \to L(x) \to X$ be the principal fibre bundle of the tangent bundles. Recall $L(X)$ is the bundle formed of the linear frames on X. Let G be a closed subgroup of $Gl(n,R)$.

### DEFINITION

A G-structure $\sigma$ on X is 'a reduction' of the structure group $Gl(n,R)$ of the bundle $Gl(n,R) \to L(X) \to X$ to the group G.

Reduction here means the following

i) there exists an open covering $U_i$ of X

ii) on each open set a coframe $w_i = (w_i, \ldots, w_i^n)$ is given such that an $U_i \cap U_j$, $w_i = g_{ij} \circ w_j$ where $g_{ij}: U_i \cap U_j \to G$. A reduction may not always be possible.

The set of all G-structures on X is in one to one correspondence with the set of all cross-sections of the reduced bundle $L(X)/G \to X$ (for proof see e.g. Kobayashi-Nomizu [475]). Hence the problem of the existence of a G-structure is a problem of the existence of cross-sections in the bundle $L(X)/G \to X$ and therefore, by Steenrod, we infer that unless certain conditions on the characteristic classes are satisfied a reduction will not be possible in general.

Without going deep into the theory of characteristic classes, we can explain why certain conditions on characteristic classes mau have to be satisfied in order that a reduction be possible. Given a manifold M and a Lie group G, there exists a bundle $E_G \to B_G$ called a universal bundle, with structure group G such that every bundle with structure group G over M is induced by a map $f: M \to B_G$. $B_G$ depends on G and the dimension of M. If $E_G \to B_G$ and $E'_G \to B'_G$ are two universal bundles, then $B_G$ and $B'_G$ are of the same homotopy type. $B_G$ is called the classifying space.

The universal bundle theorem says that: the set of all isomorphism classes of fibre bundles over M with structure group G is in one to one correspondence with the set of homotopy classes of maps of M into the classifying space $B_G$. Hence the map f induces a homomorphism $f^*: H^*(B_G) \to H^*(M)$ between the cohomology rings. The elements in the image of this homomorphism are called the characteristic classes. Suppose now $G \subset Gl(n,R)$ is a Lie subgroup. The universal bundle $E_G \to B_G$ can be considered as a $Gl(n,R)$ bundle and is hence induced by a mapping $h: B_G \to B_{Gl(n,R)}$. A bundle over M induced by a mapping $f: M \to B_{Gl(n,R)}$ is a G-bundle if and only if a mapping $f'$: $M \to B_G$ exists such that $f = h \circ f'$; that is the diagram

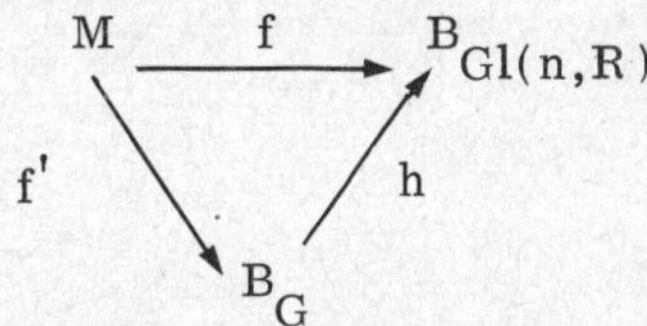


is commutative. This implies that the following diagram among the cohomology rings must also be commutative:

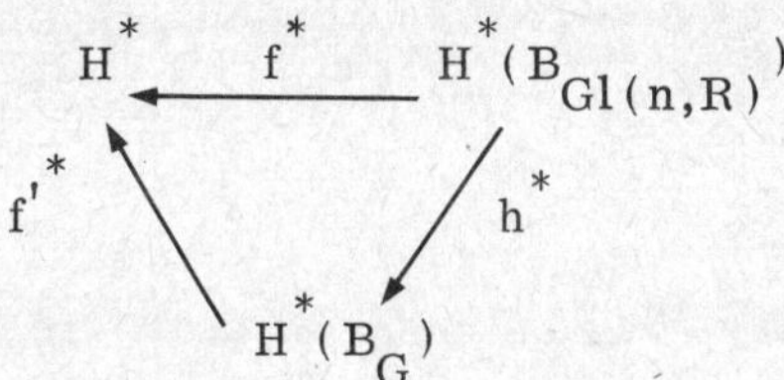


The commutativity of this diagram implies that certain relations among the characteristic classes hold in general.

## I-1-B-(11) The Integrability Problem for G-structures

Let W be a coframe on an open set U of X. We say w belongs to the G-structure $\sigma$ if $w = g_i \circ w_i$ on $U \cap U_i$ where $g_i: U \cap U_i \to G$. On $\mathbb{R}^n$, the coframe $(dx^1, \ldots dx^n)$ belongs to the standard G-structure of $\mathbb{R}^n$. A G-structure $\sigma$ on X is said to be integrable if for every $x_o \in X$, there is a local diffeomorphism $f: U \to X$ defined in a neighbourhood of the origin $\mathbb{R}^n$ such that $f(O) = x_o$ and a mapping $g: U \to G$ satisfying the partial differential equation $f^* w = g \circ \sigma$ where w is a coframe belonging to $\sigma$ on $V = f(U)$.

Let $G \subset GL(n,R)$ be a Lie subgroup. Consider the set $\Gamma_G$ of all local diffeomorphisms $f = (f^1, \ldots, f^n)$ of $R^n$ such that the Jacobson matrix $J_f = (\frac{\partial f}{\partial x^i}(x)) \in G$ for all x in the domain of f. Then $\Gamma_G$ is a pseudo group: it is in fact a Lie pseudo group. It can be shown that the set of all integrable G-structures is in one to one correspondence with the set of all $\Gamma_G$ - structures on X.

For a G-structure $\sigma$ on X to be integrable a formal integrability condition must necessarily be satisfied: given any point $x \in X$ (i) there must be a coframe w belonging to $\sigma$ on a neighbourhood of x (ii) there must exist a local diffeomorphism $f : R^n \to X$ and a mapping $g; R^n \to G$ defined in a neighbourhood of the origin in $\mathbb{R}^n$ such that $f(O) = x$ satisfying the partial differential equation $f^* w = g.\sigma$ at O (to infinite order). Then the integrability problem for the given G-structure is to show that every formally integrable G-structure is integrable.

I-1-B-(12) Examples of integrable G-structures

(a) G = the orthogonal group $O(n)$. A G-structure on X is a Riemannian metric on X. $\Gamma_G$ is the pseudo group of all isometries of the Riemannian manifold. The formal integrability condition is that the metric be flat (the curvature of X be zero). Then actual integrability of the G-structure (that is the existence of a $\Gamma_G$ structure on X, or equivalently the existence of local isometrics $f : R^n \to X$) follows from the existence theorem of differential equations.

(b) G = the symplectic group $Sp(n,R)$. Consider G as a subgroup of $GL(2n,R)$: a G-structure on X is an almost symplectic manifold structure on X. G is the pseudo group of all local symplectic diffeomorphisms of $R^{2n}$ (diffeomorphisms preserving the two form $w = dx^1 \wedge dx^{n+1} + \ldots + dx^n \wedge dx^{2n}$ of $R^{2n}$). The formal integrability condition is that $dw = O$ (i.e. the form be d-closed). If this is satisfied, then the given almost symplectic manifold is integrable (symplectic manifold) (Darboux - Frobenius). For proof of this see for example page 154, Kobayashi [473]. Hence a symplectic structure on a 2n-manifold is given by a closed 2-form of maximal rank.

(c) $G = Gl(n;C)$. Consider $Gl(n;C)$ as a subgroup of $Gl(2n;\mathbb{R})$ by associating

$$A_1 + i A_2 \to \begin{vmatrix} A_1 & A_2 \\ -A_2 & A_1 \end{vmatrix} .$$

Then $Gl(n;C)$ - the structure on X is an almost complex structure on X.

In fact there is a one-to-one correspondence between the set of all almost complex structures on a differentiable manifold X of dimension 2n and the reductions of the structure group of L(X) to Gl(n;C) (for a proof see e.g. page 142, Kobayashi-Nomizu [475]).

We have already mentioned the necessary conditions in terms of characteristic classes for the existence of almost complex structures and necessary and sufficient conditions, in some cases, in terms of the characteristic classes for the existence of almost complex structures. G is here the pseudo group of all local biholomorphic maps preserving the canonical almost complex structure. The formal integrability condition can be expressed in any one of the following three equivalent ways.

(1) The Nyjenhuis tensor of N of the almost complex structure J must vanish. N is defined to be the tensor of type (1,2) given by

$$N(Y,Z) = 2\{[JY, JZ] - [Y,Z] - J[Y,JZ] - J[JY,Z]\}$$

where Y,Z are any two differentiable vector fields on X. (N(Y,Z) is also called the torsion of J).

(2) On an almost complex manifold, the operator $\partial, \bar{\partial}$ acting on differentiable forms can be introduced (as in the case of a complex manifold). Let d be the exterior differentiation. Then the formal integrability condition is that $d = \partial + \bar{\partial}$.

(3) On an almost complex manifold we have the notions of a bundle of tangent vectors of type (O,1) and differential forms of type (O,1). Then the formal integrability condition is that there exists a differential form w of type (O,1) such that $\bar{\partial}w - \frac{1}{2}[w,w] = O$.

The actual integrability of a formally integrable almost complex structure (that is an integrable almost complex manifold is a complex manifold) was proved first by Newlander-Nirenberg [691] in 1957. This is a deep result and has also been stimulating research in subsequent years for other integrability problems. Actually Newlander-Nirenberg proved that an integrable almost complex structure of differentiability class $C^{2n+\lambda}$, $O < \lambda < 1$, is a complex structure. This was extended later to the case of differentiability class $1 + \lambda$ by Nijenhuis-Wolf [698]. Kohn [499] and Hormander [410] gave alternative proofs for the Newlander-Nirenberg theorem. The proof of the theorem becomes easier if the almost complex manifold is real analytic. (Ehresmann [227] Eckmann-Frolicher [217]). Kuranishi [514] has given a proof in the real analytic case in a slightly more general set-up.

In the case of dimension 2, the fact that an integrable almost complex structure is a complex structure follows from the existence of isothermal coordinates. Chern [153] gave an elementary proof for the existence of isothermal coordinates on a smooth surface. Finally Malgrange [563] has proved a general integrability theorem from which the Newlander-Nirenberg theorem follows. For a proof of the Newlander-Nirenberg theorem using the idea of Malgrange ('the diagonal trick') see Nirenberg [699 (2)].

(d) Let $G = Gl(n,k; \mathbb{R})$ be the subgroup of $Gl(n;R)$ consisting of all matrices leaving invariant the k-dimensional subspace of $R^n$ defined by $x_{k+1} = 0, \ldots, x_n = 0$. Then a $Gl(n,k;R)$ structure on X is a distribution F of rank k (equivalently a sub-bundle of rank k of the tangent bundle TX). $\Gamma_G$ is the pseudo group of local diffeomorphisms preserving the canonical 'foliation' of $R^n$ whose 'leaves' are the fibres of the natural projection map $\pi : R^n \to R^{n-k}$. The problem here is to find local diffeomorphisms $f : R^n \to X$ such that for all x in the domain of f, $f_*(F'_x) = F_x$ where $F'$ is the bundle of vectors tangent to the fibres of x. This gives the formal integrability condition that the space of sections of F be closed under the Lie bracket operation. If this condition is satisfied, then the distribution F is actually integrable. This can be seen by applying the Frobenius theorem (which is also referred to as the Clebsch-Deahna-Frobenius theorem).

An integrable distribution is called a Foliation. Hence a $Gl(n,k;R)$-structure on X could be called an almost foliation. More on Foliations follows in this section. For the general theory of G-structures and to see clearly how various problems of geometry can be regarded as problems of G-structures, one must read the masterly survey of Chern [154]. Even though many of the problems mentioned in the survey have now been solved, this paper will continue to guide and inspire further work for many years to come. See also Sternberg [874], Bernard [81] and Kobayashi [473]. For the local equivalence problem of G-structures see the recent lecture notes of Molino [617] and for the associated problems on exterior differential systems, especially the Cartan-Kuranishi prolongation theorem see Kuranishi [509]. (Recently Chern-Mosser [157] have solved the holomorphic equivalence problem for real hypersurfaces in $C^n$.) For the integrability problem of G-structures see Guillemin [331] and Pollack [739]. Recently from the point of view of uniformization, Kulkarni [501] has studied G-structures.

I-1-B-(13) The Spencer conjecture

The concepts of G-structures and Lie pseudo groups can be given a unified treatment by introducing "G-structures of higher degree" and associating to each transitive Lie pseudo group "a transitive filtered Lie algebra". The description of this treatment becomes rather technical and we refer the readers to Kobayashi-Nagano [474], Guillemin-Sternberg [334], Goldschmidt-Spencer [287], [288], Guillemin [332], Kumpera-Spencer [508] and Kobayashi [473]. The integrability problem for pseudo group structures (for the precise statement see, for example, Goldschmidt [286], has led to the exhaustive study of Lie equations by Malgrange, Spencer and Goldschmidt in the last few years. The whole development may be said to have started specifically to answer the Spencer conjecture that "formally integrable elliptic Lie pseudo groups are integrable" (the analogue of the Newlander-Nirenberg theorem).

This conjecture has been verified by Malgrange [563] for analytic Lie pseudo groups which are elliptic. The conjecture is not true in general, as was shown by the counter-examples of Guillemin-Sternberg [336] and Conn [170]. Recently Goldschmidt and Spencer have proved the Spencer conjecture in the affirmative for all Lie pseudo groups acting on $R^n$ which contain translations. For an account of this solution see Goldschmidt [286].

I-1-B-(14) Obstructions to the existence of an almost foliation

An almost foliation of dimension k (= a smooth distribution of rank k = an oriented smooth field of tangent k-planes) on a compact connected differentiable manifold M of dimension n is a Gl(n,k;R)-structure on M. Should M have an almost foliation of dimension k, $1 < k < n$? (The plane field problem.) If not, what are the obstructions? In general M need not have an almost foliation and the obstructions are given, as before, in terms of characteristic classes.

If a k-plane field is defined at all but a finite number of points we say that the k-plane field is with finite singularities. To measure how far is a k-plane field with finite singularities from a k-plane field without singularities, an algebraic invariant called the index can be associated. This index will be zero if and only if there is a k-plane field on M without singularities which agrees with the given k-plane field with singularities on the (m-2) skeleton of M. We state some definitive results of E. Thomas and the complete solution of the problem in dimension 4 obtained by Hirzebruch and Hopf.

Let $\xi$ be any vector bundle over M and let $w_i(\xi) \in H^i(M, Z_a)$ be its Stiefel-Whitney classes. Let $\{\xi\}$ denote the stable equivalence class of $\xi$. For any two vector bundles $\xi, \eta$ over M define $\Theta_i(\xi, \eta) = w_i((\xi) - (\eta))$, $i > 0$. Since Stiefel-Whitney classes are stable invariants, this is well defined. If $\tau$ is the tangent bundle of M, define $\Theta_i(\xi) = \Theta_i(\tau, \xi)$. The following theorems were obtained by E. Thomas [908].

THEOREM 4 (E. Thomas)

Let M be an oriented compact connected differentiable manifold of dimension n. Let $\xi$ be an oriented k-plane bundle over M, $1 < k < n$. Then $\xi$ is an almost foliation of dimension k over the (n-k+1) skeleton if and only if $\Theta_{n-k+1}(\xi) = 0$ for (n-k) odd and $\delta^* \Theta_{n-k}(\xi) = 0$ for (n-k) even.

In the above $\delta^*$ denotes the Bockstein operator corresponding to the sequence $O \to Z \to Z_2$.

More precisely we have the following theorem givng the obstructions to the existence of an almost foliation of dimension 2 in odd dimensional manifolds. We know that an oriented 2-plane bundle $\eta$ over M is determined by its Euler class $\chi(\eta) \in H^2(M, Z)$. If $\eta$ is an almost foliation of dimension 2, we call $\chi(\eta)$ its Euler class. For any $u \in H^2(M, Z)$, define $\Theta_k(u) = \sum_{2i+j=k} u^i \cup w_j(M)$ when $\cup$ denotes cup product and $w_j(M)$ are the Stiefel-Whitney classes of M. Note $\Theta_k(u) \in H^2(M, Z_2)$. Then we have

THEOREM 5 (E. Thomas)

Let M be an oriented compact connected differentiable manifold of dimension n where $n = 3 \bmod 4$. Let $u \in H^2(M, Z)$. Then M has an almost foliation of dimension 2 with Euler class u if and only if $\Theta_{n-1}(u) = O$.

For $n = 1 \bmod 4$ results of the above type are available but they are not precise.

When n is even, we have

THEOREM 6 (E. Thomas)

Let M be an n-manifold as above and assume n is even. Let $\xi$ be an oriented 2-plane bundle over M with $\delta^* \Theta_{n-2}(\xi) = 0$. Then $\xi$ is an almost foliation (with finite singularities) of dimension 2 on M.

When n = 4, this result was obtained much earlier by Hirzebruch and Hopf. In fact they completely solved the plane field problem in this case. In view of the great importance of their results, we give an account of them here.

With each oriented 2-plane bundle $\xi$ over any even dimensional manifold, one can associate an index (see for example Thomas [908] or Hirzebruch-Hopf [385]. If $V_{m,k}$ denotes the Stiefel manifold whose points are ordered k lincarly independent vectors in $R^n$, then the index $(\xi) \in \pi_{m-1}(V_{m,2})$ ((m-1)th homotopy group of $V_{m,2}$). It is known that

$$\pi_{m-1}(V_{m,2}) = \begin{cases} Z_2 & \text{if } m \text{ is odd, } m \geq 5 \\ Z \oplus Z & \text{if } m = 4 \\ Z \oplus Z_2 & \text{if } m \text{ is even and } m \geq 6 \end{cases}$$

Since $\pi_3(V_{4,2}) = Z \oplus Z$, the index of a 2-plane bundle on a compact connected oriented 4-manifold is given by a pair of integers. Let $H = H^2(M,Z)/(\text{Torsion subgroup})$. Note $H^4(M,Z) = Z$ since M is oriented. By Poincare duality, we get a symmetric bilinear nonsingular form, $S : H \otimes H \to Z$. The signature of S = (number positive eigenvalues of S) - (number of negative eigenvalues of S). The signature (or index) of M is the signature of S. It can be proved that there exists a class $w \in H$ such that $S(w,x) = S(x,x) \bmod 2$ for all $x \in H$. The set W of all w having this property is a coset H/2H. Let $\Omega = \{S(w,w) \in Z \mid w \in W\}$. If $H = 0$, take $\Omega = 0$. Then

THEOREM 7 (Hirzebruch-Hopf)

Let M be a compact connected orientable differentiable 4-manifold. Then

a) M has a field of 2 planes with finite singularities.

b) Let $\xi$ be a field of 2 planes with index $\xi$ given by the pair (a,b) of integers. Then there exist $\alpha, \beta \in \Omega$ such that

$$a = \frac{1}{4}(\alpha - 3\sigma(M) - 2\chi(M))$$

$$b = \frac{1}{4}(\beta - 3\sigma(M) + 2\chi(M))$$

where $\sigma(M)$ is the index of M and $\chi(M)$ is the Euler characteristic of M. Conversely, for any $\alpha, \beta \in \Omega$ consider the pair (a, b)of integers satisfying the above equations; then there exists a field of 2 planes with finite singularities whose index is given by this pair (a, b).

Recall that by a tangent k-field on M we mean a set of k linearly independent tangent vector fields on M. If a k-field is defined except at a finite number of points, it is said to be a k-field with finite singularities. As before we have the notion of the index of a k-field with finite singularities. The well-known theorem of Hopf that a compact manifold has a vector field without zeros if and only if the Euler characteristic is zero can be stated as "the index of a 1-field with singularities in a compact manifold M is equal to the Euler characteristic of M". The index of a 2-field with singularities will be an element of $\pi_{m-1}(V_{m,2})$. Hence when m is odd, the index is a mod 2 integer and when m is even, the index has two components; the first being called the Z-index and the second the $Z_2$-index. Then we have the following theorem on 2 fields.

THEOREM 8 (Thomas)

Let M be a compact n-manifold with $w_2(M) = 0$. Let $n = 1 \bmod 4$, $n \geq 5$. Then if $(X_1, X_2)$ is any 2-field with finite singularities, the index of $(X_1, X_2) = \hat{\chi}(M)$ as a mod 2 integer where $\hat{x}(M) = \sum_{o}^{q} (\dim H_i(M; Z_2)) \bmod 2$, $n = 2q+1$.
($\hat{\chi}(M)$ is known as the Kervaire mod 2 semi-characteristic)

THEOREM 9 (Thomas)

Let M be a compact orientable n-manifold where $n = 2 \bmod 4$, $n \geq 6$ and $(X_1, X_2)$ be any 2-field with finite singularities on M. Then the $Z_2$-index $(X_1, X_2) = 0$ as a mod 2 integer.

THEOREM 10 (Atiyah)

Let M be a compact oriented manifold of the 4k, $k > 1$. If $(X_1, X_2)$ is any 2-field with finite singularities, then the $Z_2$-index $(X_1, X_2) = \frac{1}{2}(\sigma(M) - (-1)^k \chi(M) \bmod 2$.

Atiyah has obtained a theorem on 2-fields when $n = 1 \bmod 4$ using the Atiyah-Singer index theorem without the condition $w_2(M) = 0$ in Theorem 8.

THEOREM 11 (Atiyah)

Let M be a compact orientable (4k+1) manifold with k > O. If $(X_1, X_2)$ is any 2-field with finite singularities on M, then the index $(X_1, X_2) = k(M)$ as mod 2 integers, where $k(M) = (\Sigma b_{2p}(M))$ mod 2 and $b_i$ are the Betti numbers of M. (k(M) is called the real Kervaire semi-characteristic.)

THEOREM 12 (Hirzebruch-Hopf)

Let M be as in the above theorem. Then

a) M admits a 2-field with finite singularities.

b) Let $\eta$ be a 2-field (with finite singularities) with index $\eta$ given by the pair of integers (a,b). Then there exists $\alpha \subset \Omega$ such that

$$a = \frac{1}{4}(\alpha - 3\sigma(M) - 2\chi(M))$$

$$b = \frac{1}{4}(\alpha - 3\sigma(M) + 2\chi(M)).$$

Conversely for any $\alpha \in \Omega$, consider the pair (a,b) of integers satisfying the above equations. Then there exists a 2-field with finite singularities whose index is given by the pair (a,b).

We list some important corollaries of the above theorems:

Corollaries (of Theorems 7 and 12)

1) M admits an oriented 2-plane field if and only if $(3\sigma(M) + 2\chi(M) \in \Omega$ and $3\sigma(M) - 2\chi(M) \in \Omega$.
2) M admits a 2-field if and only if $\chi(M) = 0$, $3\sigma(M) \in \Omega$
3) M is parallelizable if and only if $\chi(M) = 0$, $\sigma(M) = 0$ and $u^2 = 0$ for all $u \in H^2(M, Z)$.
4) M admits an almost complex structure if and only if $3\sigma(M) + 2\chi(M) \in \Omega$.
5) $S^4$ does not admit any almost complex structure.

Remark

For further generalizations of Hopf's index theorem see Atiyah [53], Atiyah and Dupont [58].

This is an appropriate place to mention the deep work of Adams [9] giving the solution of the famous vector field problem for spheres. The problem until 1961 was to find the maximum number of linearly independent vector fields on the sphere $S^{n-1}$.

Write any integer n in the form $n = (2\alpha-1)2^{\beta}$ where $\beta = \gamma + 4\delta$, $O \leq \gamma \leq 3$. Define $\rho(n) = 2^{\gamma} + 8\delta$. Then

THEOREM 13 (Adams)

The maximum number of linearly independent vector fields on $S^{n-1}$ is $\rho(n) - 1$.

The proof uses secondary cohomology operations introduced by Adams for K-theory (of Atiyah-Hirzebruch).

I-1-B-(15) Foliations

1. This is a vast field; major developments took place in the last decade and the field continues to be currently very active. We confine attention to only three basic amd well understood aspects of the theory of smooth foliations on a differentiable manifold (i) Topological obstructions to the integrability of an almost foliation (ii) Some existence theorems and (iii) Haefliger's classification theory. The modern theory of foliations has been built on the pioneering work of Reeb [779] and Haefliger [342].

First let us recast the definition of a foliation on a differentiable manifold in terms of local coordinates. An almost foliation F of dimension p (= a tangent p-plane field) is a smooth sub-bundle F of the tangent bundle TM of fibre dimension p. F is said to be integrable if there exist local coordinates $\{x^1, \ldots, x^p, y^1, \ldots, y^q\}$, $p+q=n$, such that F is spanned by $\{\frac{\partial}{\partial x^1}, \ldots, \frac{\partial}{\partial x^p}\}$. An integrable almost foliation is a foliation. As already remarked, integrability of F is equivalent to the condition that the set of all smooth sections of F is closed under the Lie bracket operation. This integrability condition can also be equivalently given in terms of differentiable forms. Let I(F) be the ideal of differential forms which vanish on F. Then F is integrable if and only if $d(I(F)) \subset I(F)$ where d is the exterior differentiation. If q is the codimension of F, I(F) is locally generated by q independent 1-forms $w_1, \ldots, w_q$.

The integrability condition $d(I(F)) \subset I(F)$ is equivalent to requiring that there exist a 1-form $\Theta$ such that $d(w_1 \wedge \ldots \wedge w_q) = \Theta \wedge (w_1 \wedge \ldots \wedge w_q)$. The leaves of the foliation F are minimal subsets L of M such that any smooth curve: $\alpha: [0,1] \to M$ with $\alpha(O) \in L$, $\alpha'(t) \in F$ must be such that $\alpha(1) \in L$. Equivalently, leaves are defined as equivalence classes on M as follows. Call two points x, y of M equivalent if there exists a smooth curve $\alpha: [O,1] \to M$ such that $\alpha(O) = x$; $\alpha(1) = y$ implies that $\alpha'(t) \in F$. This is an equivalence relation and the equivalence classes are called the leaves of the foliation. Let Z denote the space of all leaves of F. There is the natural projection $\pi: M \to Z$. Prescribe the quotient topology to Z. In general Z will not be Hausdorff. This is one of the difficulties one may have to face. If Z is Hausdorff, we say that the foliation itself is Hausdorff.

One of the nice properties of the Hausdorff foliation of a compact manifold is that all the leaves of any such foliation are compact. The tangent planes to the leaves of a foliation F of codimension q form a sub-bundle $\tau(F)$ of $T(M)$ of codimension q and it is integrable. $\tau(F)$ is called the tangent bundle to F and the quotient bundle $\nu(F) = T(M)/\tau(F)$ is called the normal bundle to F. Let $G^p T(M)$ be that Grassmann bundle of p-planes in $T(M)$ and let $\Gamma(G^p T(M))$ be the spacc of smooth sections of $G^p(T(M))$. $\Gamma(G^p(T(M))$ carries the structure of a Frechet manifold (of infinite dimension). Each of its elements is an almost foliation on M. The space $F^p(M)$ of all foliations of dimension p is a closed subset of $\Gamma(G^p T(M))$.

In early 1950, Reeb asked a fundamental question which was later refined by Haefliger.

The Reeb Problem

Assume that a differentiable n-manifold M admits a codimension q almost foliations $F_o$. Should M admit a foliation F of codimension q?

The Reeb-Haefliger Problem

Assume M, $F_o$ as above. Should there exist a foliation F of codimension q which is in the homotopy class of $F_o$?

Examples exist answering both the problems in the negative.

(i) The complex projective space $P^m(\mathbb{C})$, m odd and $m \geq 5$, has an almost foliation of codimension 2 but has no foliation of codimension 2 (see Bott [112]).

(ii) The manifold $P^m(\mathbb{C}) \times S^1 \times S^1$, for all m has foliations of codimension 2. If $m \geq 3$, there exists an almost foliation of codimension 2 which is not homotopic to any foliation (see e.g. Thomas [908]).

In trying to answer the above problems, it becomes necessary to differentiate between foliations of codimension 1 (dimension 1) and foliations of higher codimension (higher dimension); the theory of foliations on compact differentiable manifolds differs significantly from that on open manifolds; special care and methods are needed to treat the topological foliations. Mostly we confine attention to $C^r$-foliations, $1 \leq r \leq \infty$. We do not consider holomorphic foliations at all. Nor do we go into the general investigations of 'the internal structure' of a given foliation (i.e. the topology of foliations).

## 2. Dimension 1 and Codimension 1 Cases

It is easy to see that one-dimensional smooth foliations on a manifold are in a one-to-one correspondence with the set of smooth line fields on M. Hence a compact differentiable manifold admits a smooth one-dimensional foliation if and only if its Euler characteristic is zero. Every open manifold (no component being compact) does admit a one-dimensional foliation. It was conjectured by Emery Thomas that a compact differentiable manifold admits a codimension one foliation if and only if its Euler characteristic is zero. This has been proved to be correct (see e.g. Lawson [529]). It was proved by Philips [732] that on an open manifold every codimension one-plane field is homotopic to a smooth foliation. Is this true over a compact manifold? That has been proved to be true for a transversely orientable codimension one-plane field on compact manifolds (Wood-Schweitzer-Thurston theorem). In lower dimensional cases we have more precise results: every compact 3-manifold has a smooth codimension one foliation (Alexander-Lickovish-Novikov-Reeb-Wood); every odd dimensional sphere $S^{2n+1}$ admits a smooth codimension one foliation (Durfee-Tamura); every odd dimensional exotic sphere $\Sigma^{2n+1}$ admits a smooth codimension one foliation (Lawson-Milnor). The references for these results is the book by Lawson [529], and the papers by Durfee [214], Lawson [528], Tamura [889].

One must emphasize here the difference between smooth foliations and real analytic foliations. A famous 'non-existence' theorem of Haefliger asserts that no simply connected compact differentiable manifold can admit any codimension one real analytic foliations. The classifying space (refer to 5 in this section) of codimension 1 real analytic foliations has the type of the Eilenberg Maclane space $K(\pi, 1)$, where $\pi$ is an uncountable torsion free perfect group (Haefliger).

Reeb was the first to construct a codimension one foliation on $S^3$. This is an ingenious construction. $S^3$ can be viewed as the union of two solid tori intersecting in a common toroidal surface. Reeb constructed in the interior of each of these solid tori a $C^\infty$ codimension 1 foliation. These two foliations together with the common surface patch up to give a $C^\infty$ codimension 1 foliation on $S^3$. First he constructed a particular codimension 1 foliation on $\mathbb{R}^2$ as follows: consider real valued $C^\infty$ functions $f(x)$ defined in $(-1,1)$ such that $\lim_{|x|\to 1} f^{(k)}(x) = \infty$. The graphs of the functions $y = f(x) + c$, $c \in R$ and the vertical lines $x = \alpha$, $|\alpha| \geq 1$ foliate $R^2$. Then by rotating the strip $\{(x,y) \in R^2 \mid -1 < x < 1\}$ around the y axis in 3-dimensions he obtained a codimension $1(C^\infty)$ foliation of $\dot{D}^2 \times R$ where $\dot{D}^2$ denotes the interior of the 2-disk. This foliation is invariant under translations in the vertical direction and gives the Reeb foliation.

Another result that demands a place here is the Godbillon-Vey invariant. Let F be a codimension $1(C^\infty)$ foliation of a compact manifold defined by a completely integrable 1-form and which is a 1-form w such that $w \wedge dw = 0$. Take any 1-form $\Theta$ such that $dw = -\Theta \wedge w$. Then Godbillon-Vey [277] proved that the de Rham cohomology class $\in H^3(M,R)$ of the closed form $\Theta \wedge d\Theta$ is an invariant of the given foliation. This cute result has led to further significant invariants for foliations.

## 3. Topological Obstructions to Integrability

In 1968 Bott [112] obtained a topological obstruction to integrability. This is a remarkable discovery and led to a lot of significant developments in the theory of foliations.

THEOREM 14 (Bott's Vanishing Theorem)

Let F be a foliation of codimension q, on a differentiable manifold M, with normal bundle $\nu(F)$. Then $\mathrm{Pont}^{2k}(\nu(F)) = 0$ for all $k > q$, where $\mathrm{Pont}^*(\nu(F))$ denotes the subring of $H^*(M,\mathbb{R})$ generated by the Pontrjagin classes of $\nu(F)$.

Surprisingly the proof of this theorem is not that difficult. Bott constructed a special connection (now known as the Bott connection or the basic connection) on $\nu(F)$ which is flat along the leaves; then a straightforward application of the classical Weil homomorphisms gives the above result. Note that the condition that $\mathrm{Pont}^{2k}(\nu(F)) = 0$, $k > q$, depends only on the homotopy class of $\nu(F)$. It is important to observe here that this is only a primary obstruction to integrability and it is not true when R is replaced by Z or any finite field. Schulman [818] showed in his thesis in 1972 that there are secondary obstructions to integrability which exist independently of the primary obstructions given by Bott: let $\alpha, \beta, \gamma$ be closed differential forms on

M of degree a,b,c respectively and let $\alpha \wedge \beta = dp$, $\beta \wedge \gamma = d\sigma$. Then their Massey triple product $< \alpha, \beta, \gamma >$ is the closed differential form of degree (a+b+c-1) given by $< \alpha, \beta, \gamma > = \alpha \wedge \sigma + (-1)^{a+1} p \wedge \gamma$. Its cohomology class $[\alpha, \beta, \gamma]$ is well defined in

$$H^{a+b+c-1}(M,R) / H^{a}(M,R) \cup H^{b+c-1}(M,R) + H^{c}(M,R) \cup H^{a+b-1}(M,R)$$

It is known that Massey triple products are homotopy invariants. Now we can state Schulman's result.

THEOREM 15 (Schulman)

Let F be a foliation of dimension q on a manifold M with normal bundle $\nu(F)$. Then for all $\alpha, \beta, \gamma$ in $\mathrm{Pont}^*((F))$ with degree $(\alpha.\beta) > 2q$ and degree $(\beta, \gamma) > 2q$, the Massey triple product $[\alpha, \beta, \gamma] = 0$.

## 4. Foliations are Generalized Principal Bundles

A foliation of dimension zero on M is just the differentiable structure of M and the leaves are just the points of M. Hence one way of looking at foliations is to regard them as generalized differentiable structures. Also note that if $f: M \to N$ is any submersion of differentiable manifolds, then f induces a natural foliation of codimension n (= dim N) on M. The leaves of this foliation are the components of $f^{-1}(y)$, $y \in N$. The fact that every foliation on a manifold M is given locally by submersions is inherent in the definition of foliations given earlier. The following definition explicitly brings out the two facts that foliations are generalized differentiable structures and are locally given by submersions.

DEFINITION

A smooth foliation of codimension q on a differentiable manifold M is a maximal family of differentiable submersions $f_\alpha : U_\alpha \to R^q$ where $\{U_\alpha\}$ $\alpha \in \Lambda$ is an open cover of M and for each $\alpha, \beta \in \Lambda$, each $x \in U_\alpha \cap U_\beta$, there exists a local diffeomorphism $\varphi^x_{\beta\alpha}$ of $R^q$ such that $f_\beta = \varphi^x f_\alpha$ holds in a neighbourhood of $U_x$ of x.

The transition functions $\varphi^x_{\beta\alpha}$ satisfy

(i) compatibility condition: $\varphi^x_{\beta\alpha} = \varphi^y_{\beta\alpha}$ on $f_\alpha(U_x \cap U_y)$ and

(ii) cocycle condition: $\varphi^x_{\beta\alpha} = \varphi^x_{\beta\gamma} \varphi^x_{\gamma\alpha}$ for $x \in U_\alpha \cap U_\beta \cap U_\gamma$.

Let $\Gamma_q$ be the set of all germs of local diffeomorphisms of $R^q$. This is a typical example of a groupoid (and not a group in general). The units in $\Gamma_q$ are the germs of the identity maps at points $x \in R^q$. Let $\sigma_o, \sigma_1: \Gamma_q \to R^q$ be the maps which map a germ to its source and target respectively. Each $\varphi \in \Gamma_q$ has an inverse, $\varphi^{-1}$, such that $\varphi^{-1}\varphi = 1_{\sigma_o(\varphi)}, \varphi.\varphi^{-1} = 1_{\sigma_1(\varphi)}$. $\Gamma_q$ is further a topological groupoid: it has a topology for which the composition and inverse are continuous. There is a natural topology on $\Gamma_q$ having a basis consisting of sets of the form $\{\underline{\varphi}_x : x \in \text{domain } \varphi\}$ where $\varphi$ is a local diffeomorphism of $R^q$ and $\underline{\varphi}_x$ denotes the germ of $\varphi$ at x. Any topological group is trivially a topological groupoid. A pseudo group G whose elements are local homeomorphisms of open sets of any topological space gives rise to a unique topological groupoid, formed by the germs of the maps $f \in G$ at the points of the space.

In view of the above remarks on $\Gamma_q$ the definition given above shows that a foliation can be regarded as a generalized principal bundle over M with fibre being a topological groupoid (recall that for the usual principal bundles the fibres are topological groups). Hence it is natural to look for classifying spaces for foliations. In fact, in a more general set-up, Haefliger established the existence of classifying spaces and studied their properties and related these to foliations.

## 5. Haefliger structures

Let X be a differentiable n-manifold. A smooth $\Gamma_q$ cocycle on X, $q \leq n$, is given by a collection $(U_\alpha, f_\alpha)$ $\alpha \in \Lambda$ where $U = \{U_\alpha\}$ form an open covering of X, $f_\alpha$: $U_\alpha \to R^q$ are smooth maps (called local projections) such that for each $\alpha, \beta$; $x \in U_\alpha \cap U_\beta$, there exists an element $\varphi^x_{\beta\alpha} \in \Gamma_q$ such that (i) $f_\beta = \varphi^x_{\beta\alpha}$ holds in a neighbourhood of x; and (ii) for each $x \in U_\alpha \cap U_\beta \cap U_\gamma$, $\varphi^x_{\gamma\beta} = \varphi^x_{\gamma\beta} \cdot \varphi^x_{\beta\alpha}$ holds in a neighbourhood of $f_\alpha(x)$ (cocycle condition).

Two such cocycles $\{U_\alpha, f_\alpha, \varphi^x_{\beta\alpha}\}$ $\alpha, \beta \in I$, $\{U'_\alpha, f'_\alpha, \varphi^{x'}_{\beta\alpha}\}$ $\alpha, \beta \in J$ are called equivalent if they extend to a cocycle on the disjoint union of the coverings. An equivalence class of $\Gamma_q$ cocycles is called a codimension q Haefliger structure $H_q$ on X. $H_q$ is denoted by $H_q = (u_\alpha, f_\alpha, \varphi^x_{\beta\alpha})$. If $H_q$ is a Haefliger structure on X and if $f: Y \to X$ is a differentiable map between differentiable manifolds Y and X, then this map induces a natural Haefliger structure on Y given by $\{f^{-1}U_\alpha, f_\alpha \circ f, \varphi^{f(x)}_{\beta\alpha}\}$ which is denoted by $f^* H_q$ and is called the pull back of $H_q$. If $H_1$ and $H_2$ are two Haefliger structures of codimension q on X, they are said to be homotopic (concordant) if there exists a codimension q Haefliger structure H on $X \times [0,1]$ such that $H_j = i_j^* H$ where $i_j : X \to X \times [0,1]$ are the natural inclusions $i_j(x) = (x,j)$ for $j = 0,1$. The normal bundle $\nu(H_q)$ to a Haefliger structure $H_q$ of codimension q on X is defined to be the q-dimensional vector bundle over X given by the transition functions $g_{\beta\alpha} = d\,\varphi^x_{\beta\alpha}$ in a neighbourhood of x. It is easy to check that homotopic (concordant) Haefliger structures have isomorphic normal bundles.

We observe the following important facts relating the newly introduced Haefliger structures and the foliations structures.

(1) Let $H_q = (U_\alpha, f_\alpha, \varphi^x_{\beta\alpha})$ be a Haefliger structure of codimension q on X. If all the local projections $f_\alpha$'s are submersions, then $H_q$ is a codimension q-foliation as follows from definition. In this case the $\varphi^x_{\beta\alpha}$'s are completely determined by $f_\alpha$ and the cocycle condition is automatically satisfied. Further the normal bundle $\nu(H_q)$ of the Haefliger structure $H_q$ is the same as (isomorphic) the normal bundle $\nu(H_q)$ of the foliation $H_q$. Thus Haefliger structures are more general than foliations.

(2) In the definition of a Haefliger structure instead of taking the specific topological groupoid, $\Gamma_q$, we take an arbitrary topological groupoid $\Gamma$, we get the notion of a $\Gamma$-structure over X.

(3) If $\Gamma$ happens to be a topolological group, then a $\Gamma$- structure is nothing but a principal $\Gamma$-bundle over X.

(4) Also observe that associated to any pseudo group structure (as defined earlier in the previous section) there is a $\Gamma$-structure in the present sense, $\Gamma$ being interpreted as the topological groupoid formed of the germs of the elements of the pseudo group $\Gamma$.

(5) Instead of requiring smoothness, we could easily formulate Haegliger structures $\Gamma^r_q$ of less differentiability r, $0 \leq r < \infty$. We have denoted $\Gamma^\infty_q$ by $\Gamma_q$. Also we could formulate the notion of real analytic Haefliger structures $\Gamma^\omega_q$ on X.

We now state Haefliger's classifying space theorem in this general set-up.

THEOREM 16 (Haefliger)

Let $\Gamma$ be a topological groupoid. There exists a classifying space $B\,\Gamma$ for the

$\Gamma$-structures; that is there exists a topological space $B\Gamma$ equipped with a $\Gamma$-structure $H_\Gamma$ (called a universal $\Gamma$-structure), such that

(1) if H is any $\Gamma$-structure on a paracompact Hausdorff topological space X, then there is a continuous map $f: X \to B\Gamma$ such that H is the pull back of $H_\Gamma$.

(2) If $H_1, H_2$ are two $\Gamma$-structures on X, these are concordant if and only if the associated maps are homotopic. There is a one-to-one correspondence between the concordant classes of $\Gamma$-structures on X and the homotopy classes of maps $X \to B\Gamma$.

(3) if $h: \Gamma \to \Gamma'$ is a homomorphism of topological groupoids, then there is a continuous map from $B\Gamma \to B\Gamma'$.

This is a generalization of the classical theorem on the existence of classifying spaces $BGL_q(R)$ for q-vector (principal) bundles.

In particular Haefliger structures of codimension q on a smooth manifold X have classifying spaces. Let $X_{\Gamma_q^r}$ be the universal Haefliger structure on X. The normal bundle to this structure is classified by a map $d: B\Gamma_q^r \to BGL_q$. Let $\nu$ be a q-dimensional vector bundle over X with classifying map $f_\nu: X \to BGL_q$. Suppose there exists a lifting of f over d, that is making the diagram commutative.

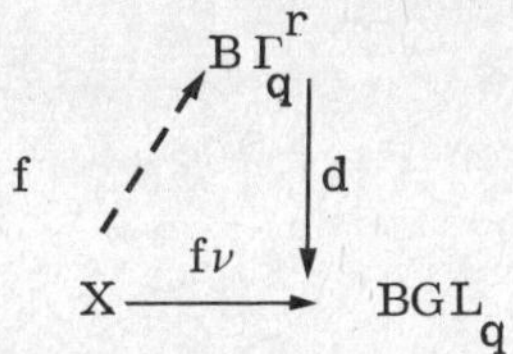


Then $\nu$ will be the normal bundle to an $H_q^r$ structure on X and conversely.

Let $B\bar{\Gamma}_q^r$ be the classifying space for an $H_q^r$-structure with a trivialized normal bundle. Then it can be shown that the obstructions to lifting all lie in $H^k(X; \pi_{k-1}(B\bar{\Gamma}_q^r))$. Hence it becomes necessary to understand the homotopy type of these spaces $B\Gamma_q^r$ and $B\bar{\Gamma}_q^r$ There are a number of deep results known regarding these. For example $\pi_i(B\bar{\Gamma}_q^r) = 0$ for $1 \leq r \leq \infty$. See in particular Haefliger [345], Mather [577],Thurston [910], Bott-Haefliger [113].

Now let us assume that the Haefliger structures $H_q^r$ are $C^r$-Foliations of codimension q on X. $B\Gamma_q^r$ is the classifying space for the foliations. The map $d: B\Gamma_q^r \to BGL_q(R)$ induces a map at the cohomology level. From Bott's theorem we can get

THEOREM 17 (Bott)

$d^*: H^k(BGL_q(R), \mathbb{R}) \to H^k(B\Gamma_q^r, R)$ is zero for all $k > 2q$ and $r \geq 2$.

From the theorem of Haefliger-Mather-Thurston, it can be proved that

THEOREM 18 (Haefliger-Mather-Thurston)

$d^*: H^k(BGL_q(R)) \approx H^k(B\Gamma_q^r)$ for $O \leq k \leq q+1$, the coefficient ring of the cohomology being arbitrary.

Bott's theorem gives rise to certain new (exotic) characteristic classes and to real cohomology on $B\Gamma_q^r$. These are closely related to the Gelfand-Fuks-cohomology of formal vector fields in $R^q$ and to 'the continuous cohomology' of $\Gamma_q^r$. Deep results have been obtained in this direction in the last few years. It is impossible here to summarize these developments. We refer the readers to the original references in addition to the references already cited: Chern-Simmon [158], Gelfand-Fuks [268], Gelfand [267], Godbillon-Vey [277], Guillemin [333], Haefliger [345], Kamber-Tandeur [439], Pittie [734].

## 6. Haefliger Structures and Foliations

Let $F_o, F_1$ be two codimension q foliations on a differentiable open manifold M. They are said to be integrably homotopic if there is a codimension q foliation $F_t$ on $M \times [O,1]$ which is transverse to $M \times \{t\}$ for all t and $F_o \approx F_o$, $F_1 \approx F_1$ (if M is compact, integrable homotopy is equivalent to the stronger notion of isotopy). An integrable homotopy class of $C^r$ foliations on M of codimension q determines a Haefliger structure $H_q^r$ and a unique (up to homotopy) embedding of $\nu(H_q^r)$ into T(M). A concordance class of Haefliger structures $[H_q]$ on M is said to be tangentially embedded if there exists a homotopy class of bundle embeddings $\nu([H_q]) \to T(M)$.

In the case of compact manifolds a concordance class of Haefliger structures $H_q$ is said to be tangentially concordance class if there exists a concordance class of bundle embedding $\nu(H) \to T(M)$. Then we have the following important deep theorems relating Haefliger structures and Foliations.

THEOREM 19 (Gromov-Philips-Haefliger-Hausmann)

Let M be an open $C^r$-manifold. $1 \leq r \leq \infty$. Then for each $k \leq r$ and $q = O, \ldots,$ $(m-1)$ the set of tangentially embedded concordance classes $[H_q^r]$ of codimension q

Haefliger $C^k$-structures on M is in one-to-one correspondence with the set of all integrable homotopic classes of codimension q, $C^k$ foliations on M.

THEOREM 20 (Thurston)

Let M be a compact $C^r$ m-manifold, $0 \leq r \leq \infty$. Then for each $k \leq r$, $q = 0, \ldots, (m-1)$, the set of tangentially concordance classes of codimension q, $C^k$ Haefliger structures ($H^r_q$) is in one-to-one correspondence with the set of all concordance classes of codimension q, $C^k$ foliations on M.

From the first theorem, we get that a necessary and sufficient condition for an almost foliation F of codimension q on an open manifold to be homotopic to a foliation is that there exists a codimension q Haefliger structure $H_q$ and a vector bundle isomorphism of $TM/F \approx \nu(H_q)$ where $\nu(H_q)$ is the normal bundle of $H_q$.

Thurston's theorem is a very deep theorem. Its proof is far more difficult and qualitatively different from the proof of the corresponding classification theorem given above for foliations on open manifolds.

We summarize the entire discussion of Part B by the following sequence of implications: (OE Diff = oriented even dimensional differentiable manifolds).

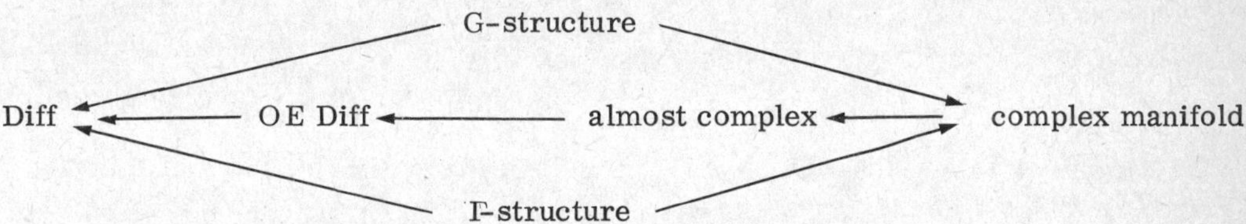


Section 1: Part C

## I-1-C: Complex Analytic Spaces, Varieties, Schemes and Algebraic Spaces

### I-1-C-(1) Complex analytic sets

We assume familiarity with the fundamental notions of sheaf theory. A good reference to this subsection is Gunning-Rossi [339].

Let $\Theta_z$ be the set of germs of holomorphic functions at $z \in C^n$. It is easy to check the following facts:

$\Theta_z$ is a commutative ring with identity; it is isomorphic with a ring of convergent power series centred at the point; it is an integral domain (its quotient field is called the field of germs of meromorphic functions at z); it is a local ring, the non-units forming the unique maximal ideal; it is also Noetherian; every ideal in $\Theta_z$ has a finite basis; and it is further a unique factorization domain. For local study it suffices to consider only the ring $\Theta_o$ at the origin O. Let

$$\Theta = \bigcup_{z \in \mathbb{C}^n} \Theta_z$$

We define a topology on $\Theta$ as follows. An element of $\Theta$ is a germ $f_b$ at some point $b \in \mathbb{C}^n$. Let $(U,f)$ define $f_b$. Let $N(U,f) = \{f_c \mid c \in U\}$ where $f_c$ is the germ at c defined by $(U,f)$. We define the sets $N(U,f)$, as $(U,f)$ runs through all the pairs defining $f_a$ to be the fundamental systems of neighbourhoods of $f_a$. We have the natural projection map $p: \Theta \to \mathbb{C}^n$ defined by $p(f_a) = a$. One can check that p is continuous, that the topology on $\Theta$ is Hausdorff and that p is a local homeomorphism. Moreover letting

$$\Theta \times_p \Theta = \{ (x,y) \in \Theta \times \Theta \mid p(x) = p(y) \}$$

we can check that the maps

$$\Theta \times_p \Theta \to \Theta \text{ defined by } (x,y) \to x - y, \ (x,y) \to xy$$

are continuous in the topology of $\Theta$. $\Theta$ is a typical example of a sheaf of rings over $\mathbb{C}^n$. $\Theta$ is called the sheaf of germs of holomorphic functions on $\mathbb{C}^n$ and is also denoted by $\Theta_{hol}$. We have three other sheaves of rings on $\mathbb{C}^n$: $\Theta_{cont}$, the sheaf of germs of continuous complex valued functions ; $\Theta_{diff}$, the sheaf of germs of differentiable complex valued functions and $\Theta_{alg}$, the sheaf of germs of rational (quotient of polynomials locally) functions on $\mathbb{C}^n$. Sections of $\Theta_{hol}$ (respectively $\Theta_{alg}$, $\Theta_{diff}$, $\Theta_{cont}$) over an open set $U \subset \mathbb{C}^n$ are the holomorphic (respectively, the rational, the complex valued differentiable, the complex valued continuous) functions on U.

We have the following sequence of implications

$$\Theta_{alg} \subset \Theta_{hol} \subset \Theta_{diff} \subset \Theta_{cont}.$$

$(\mathbb{C}^n, \Theta_{alg})$, $(\mathbb{C}^n, \Theta_{hol})$, $(\mathbb{C}^n, \Theta_{diff})$, $(\mathbb{C}^n, \Theta_{cont})$ are all typical examples of ringed spaces. A ringed space $(X, \Theta_x)$ consists of a Hausdorff topological space X and a subsheaf $\Theta_x$ of the sheaf of continuous complex valued functions on X. A morphism between two ringed spaces $(X,\Theta_x)$ and $(Y,\Theta_y)$ is a continuous map $\varphi: X \to Y$ such that for every open set $U \subset Y$, $f \in \Gamma(U, \Theta_y)$ (: = sets of sections of $\Theta_y$ over U), $f \circ \varphi \in \Gamma(\varphi^{-1}(U), \Theta_x)$. An isomorphism of a ringed space $(X,\Theta_x)$, $(Y,\Theta_y)$ is a homeomorphism $\varphi: X \to Y$ such that $\Gamma(\varphi^{-1}(U), \Theta_x) = (U, \Theta_y)$ for all open sets U of Y. Ringed spaces form a category.

One of the important concepts in the function theory of several complex variables is the concept of a coherent analytic sheaf. An analytic sheaf over a domain D in $\mathbb{C}^n$ is a sheaf of $\Theta$-modules over D.

DEFINITION

An analytic sheaf S over an open set D in $\mathbb{C}^n$ is said to be coherent if in some open neighbourhood $U_a$ of each point a in D there is an exact sequence of analytic sheaves over $U_a$ of the form

$$(O_D | U_a)^p \to (O_D | U_a)^q \to (S | U_a) \to O$$

for some positive integers p and q.

Note that coherence is a local property. When an analytic sheaf is coherent it has nice properties. For example, a general theorem of Cartan-Serre says that the cohomology groups of a compact complex space with coefficients in a coherent analytic sheaf are finite-dimensional vector spaces. But some of the difficult theorems in several complex variables are in proving that certain analytic sheaves are coherent. It is evident from the definition that the structure sheaf $O_D$ of an open set D in $\mathbb{C}^n$, considered as an analytic sheaf over $O_D$ itself, is coherent.

The following are two important theorems on coherence:

THEOREM 1 (Oka [709])

For any analytic sheaf homomorphism $\phi: O_D^r \to O_D^s$, the kernel and the image are coherent analytic sheaves.

THEOREM 2 (Serre [830])

For any exact sequence of analytic sheaves of the form

$$O \to R \to S \to T \to O,$$

if any two of them are coherent so is the third.

An important coherent sheaf is the ideal sheaf of analytic set. We now define this concept, after giving some basic facts on complex analytic sets. (By an analytic set we mean a complex analytic set.)

DEFINITION

An analytic set A in an open set D is subset A of D having the following property: to each point $a \in A$, there is a neighbourhood $U_a$ in D and holomorphic functions $f_1, f_2, \dots, f_r$ in $U_a$ such that $U_a \cap A = \{x \in U_a \mid f(x) = 0 = f_2(x) = \dots = f_r(x)\}$.

It can be proved that the set of common zeros of a collection of holomorphic functions defined in an open set is an analytic subset. Consider the set of pairs $(A_1, D_1)$ where $D_1$ is an open neighbourhood of the origin in $\mathbb{C}^n$ and $A_1$ is an analytic subset of $D_1$. Two such pairs $(A_1, D_2)$ and $(A_2, D_2)$ are called equivalent if there is an open neighbourhood W of the origin, $W \subseteq D_1 \cap D_2$, such that $W \cap A_1 = W \cap A_2$. This is an equivalence relation, and an equivalence class is called a germ of an analytic set at the origin in $\mathbb{C}^n$. For brevity, we call a germ of an analytic set an analytic germ and denote the germ of an analytic set A at the origin by $A_o$.

Let $f_o$ be the germ of the holomorphic function f at the origin. Let Z(f) be the zero set of f. If $Z(f)_o \supseteq A_o$, we say that f vanishes on A. Define id(A) = $\{f \in \Theta_o \mid f$ vanishes on A$\}$. id(A) is an ideal in $\Theta_o$. Thus to each analytic germ $A_o$ there is a canonically associated ideal in $\Theta_o$, called the ideal of A. Now, given an ideal I in $\Theta_o$, we can associate an analytic germ at the origin, called its locus, as follows: loc(I) = $Z(g_1, \dots, g_r)_o$ where $g_1, \dots, g_r$ are in $\Theta_o$ and generate the ideal I. This is well defined.

The most important relation connecting an ideal I and its locus is the Hilbert basis theorem: id loc I = radical of I. This is one of the basic theorems in the local theory of several complex variables. Using the fact that $\Theta_o$ is Noetherian, we can prove that any analytic germ can be uniquely written as an irredundant union of finitely many irreducible analytic germs. We call an analytic germ irreducible if and only if it cannot be expressed as the union of two proper subanalytic germs. It is easy to check that an analytic germ is irreducible if and only if its ideal is prime.

A set of co-ordinates $Z_1, \dots, Z_n$ at the origin in $\mathbb{C}^n$ is called a regular system of co-ordinates for a prime ideal $I \subseteq {}_nO_o$ if for some integer k, $0 \leq k < n$, we have (i) ${}_kO_o \cap I = 0$ and (ii) ${}_{j-1}O_o[Z_j] \cap I$ contains a Weierstrass polynomial in $Z_j$ for $j = k+1, \dots, n$. It is a basic fact that every prime ideal has a regular system of co-ordinates (e.g. [339], p. 93). The integral k is called the dimension of the ideal. The dimension of a prime ideal depends not only on the ideal but also on the choice of co-ordinates. However, it can be proved that if there is a regular system of co-ordinates for a prime ideal with respect to which the ideal has dimension k, then a dense open subset of a sufficiently small neighbourhood of the origin in any analytic set representing the germ of its locus is a k-dimensional complex manifold. Hence we can speak of the dimension of a prime ideal. For any analytic germ A, its dimension is defined as the maximum of the dimensions of its irreducible components. If all the irreducible components of an analytic germ have the same dimension, then the analytic germ is said to have pure dimension.

Let A be an analytic set in a domain D in $C^n$. A point $a \in A$ is called a regular point of A of dimension p if there exists a neighbourhood $U_a$ of a in D, such that $A \cap U_a$ is a complex sub-manifold of $U_a$ of dimension p. It follows that $a \in A$ is a regular point of dimension p if and only if there exist functions $f_{p+1}, \ldots, f_n \in O_a$ such that in a neighbourhood of a, $A = \{x \mid f_{p+1}(x) = 0 = f_{p+2}(x) = \ldots = f_n(x)\}$ and $(df_{p+1})_a, \ldots, (df_n)_a$ are linearly independent. A point $a \in A$ is called a singular point if it is not a regular point.

One of the difficulties with singularities is that a complex analytic set can never be a smooth manifold throughout a neighbourhood of a singular point (Milnor [60], p. 13). But there are complex analytic sets which are topological manifolds (Brieskorn [118] around each singularity. There are also examples of complex analytic sets X with singularities around which X is not even a topological manifold (see e.g. Mumford [641], pages 12 to 14). Regarding the structure of regular and singular points we have the following

THEOREM 3

Let A be an analytic set in a domain D in $\mathbb{C}^n$. Then

(a) Let $a \in A$ and dimension $\underline{A}_a$ be p. Then any neighbourhood of a contains points at which A is regular of dimension p. In particular, the set of regular points of A is dense in A.

(b) The set of singular points of A is an analytic set in D.

Let A be an analytic subset in a domain D in $\mathbb{C}^n$. To every $z \in D$, associate the ideal $id(A_z)$ and write

$$g(A) = \bigcup_{z \in D} id(\underline{A}_z).$$

Then g(A) is an open subset of the ring $\Theta_D$, and it is a sheaf of ideals on D. It is called the sheaf of ideals of the analytic set A. It is known that g(A) is coherent (see e.g. [339], p. 138). Hence the quotient sheaf ${}_A\Theta_D = \Theta_D/g(A)$ is a coherent analytic sheaf. Since g(A) is a sheaf of ideals, ${}_A\Theta_D$ is actually a sheaf of rings with support A. Let ${}_A\Theta = {}_A\Theta_D|A$. ${}_A\Theta$ is called the sheaf of germs of holomorphic functions on A.

This definition is motivated by the following. Let f be a continuous complex-valued function on A. f is said to be holomorphic at $z \in A$ if there exists an open neighbourhood W of z in D and a holomorphic function F in W such that $F|W \cap A = f|W \cap A$. Then one has the notion of the rings of germs of holomorphic functions on A and the sheaf $\Theta_A$ of holomorphic functions on A. It can be checked that ${}_A\Theta = \Theta_A$. We take the pairs $(A, \Theta_A)$ to be the local models for reduced analytic spaces.

I-1-C-(2) Complex analytic spaces

DEFINITION

A reduced complex analytic space (complex analytic space in the sense of Cartan-Serre) is a pair $(X, \Theta_X)$ where X is a Hausdorff space and $\Theta_X$ is a sheaf of subalgebras of the sheaf of germs of continuous functions, which is locally isomorphic to a local model, i.e. for each $x \in X$ there is a neighbourhood U of x in X, a local model $(A, \Theta_A)$ and a homeomorphism $\phi: U \to A$ with the property that for $y \in U$, $f \in C_{u,y}$ belongs to $\Theta_{u,y}$ if and only if $f = g \circ \phi$ for some germs $g \in \Theta_{A,\phi(y)}$.

A morphism between one reduced complex space $(X, \Theta_X)$ into another $(Y, \Theta_Y)$ is a continuous map $f: X \to Y$ such that $f^*(\Theta_{Y,f(x)}) \subset \Theta_{X,x}$ for all $x \in A$. Note that complex manifolds are special types of reduced complex space.

To define non-reduced complex spaces, we define their local models first.

Let D be open in $\mathbb{C}^n$, and let I be any coherent sheaf of ideals in $\Theta_D$. Let V = Supp $(\Theta_D/I)$. Then V is an analytic set in D. The restriction $\Theta_D/I$ to V is denoted by $\Theta_V$. In general it is not a subsheaf of $C_V$. We define $(V, \Theta_V)$ to be a local model of non-reduced type. It is a local model of reduced type if and only if I is the sheaf of all germs of holomorphic functions vanishing on V. In the non-reduced case the local model $(V,\Theta_V)$ is not determined by V alone; the structure sheaf also has to be specified. For an example of a non-reduced model, take $D = \mathbb{C}$, I being the sheaf of ideals generated by $Z^2$, where Z denotes the co-ordinate function in $\mathbb{C}$. Then $V = \mathrm{Supp}\,\Theta_{\mathbb{C}}/(Z^2) = \{0\}$ and $\Theta_{V,o} = \mathbb{C}\{Z\}/(Z^2)$ is the space of dual numbers represented by $a + b\epsilon$, where $a,b \in \mathbb{C}$ and $\epsilon^2 = 0$. Certainly $\Theta_{V,o}$ cannot be a subring of the ring of continuous functions on $\{O\}$.

DEFINITION

A non-reduced complex analytic space (complex analytic space in the sense of Grothendieck) is a pair $(X,\Theta_X)$, where X is a Hausdorff topological space and $\Theta_X$ is a sheaf of local $\mathbb{C}$-algebras, which is locally isomorphic to a local model of non-reduced type.

THEOREM 4

A non-reduced complex space $(X,\Theta_X)$ is a reduced complex space if and only if the local rings $\Theta_{X,x}$ have no nilpotent elements for any $x \in X$.

Given a non-reduced complex space $(X,\Theta_X)$, we can associate to it a reduced complex space as follows. Let $N_x$ be the ideal in $\Theta_{X,x}$ consisting of all nilpotent elements. Then

$$N = \bigcup_{x \in X} N_x$$

is a coherent sheaf: for in a local model $(V,\Theta_V)$ for $(X,\Theta_X)$ we have $N_X = (I'/I)_X$ where $I'$ is the sheaf of germs of holomorphic functions vanishing on V, and I is the sheaf of ideals defining the structure sheaf $\Theta_V$. The sheaf $I'$ is coherent (Oka-Cartan) and I is coherent by our assumption. Hence the quotient sheaf $I'/I$ is coherent. Now define $(X_{red},\Theta_{Xred})$ by $X_{red} = X$ and $\Theta_{Xred} = \Theta_X/N$.

One of the significant differences between reduced and non-reduced complex space is that decomposition into irreducible components has no meaning, for a non-reduced complex space.

The Zariski tangent space at a point x of any complex space is defined to be the $\mathbb{C}$-dual of $(m_x \mid m_x^2)$ where $m_x$ denotes the maximal ideal of $\Theta_{X,x}$. If X is defined by the ideal $I \subset \Theta_D$, D open in $\mathbb{C}^n$, the tangent space may be identified with the linear variety defined by the linear parts of all germs $\in I_x$. The Zariski tangent space of $X_{red}$ may be strictly contained in that of X.

A coherent analytic sheaf of any complex space $(X, \Theta_X)$ is a sheaf S of $\Theta_X$-modules such that for each $x \in X$ there exists an exact sequence of the form

$$(\Theta_X | U)^p \to (\Theta_X | U)^q \to (S | U) \to O$$

for some positive integers p and q.

Let $(X, \Theta_X)$ be any complex space and S be any coherent sheaf of $\Theta_X$-modules. The cohomology groups $H^q(X, S)$ are well defined, with a topology of a quotient of a Frechet space, and hence in general topology is not separated. If X is compact, we have the following finiteness theorem:

THEOREM 5 (Cartan-Serre, see e.g. [562])

Let X be any compact space and S a coherent analytic sheaf on X. Then for every $q \geq O$ the cohomology groups $H^q(X, S)$ are separated and finite dimensional.

Let $(X, \Theta_X)$ be any complex space and $x \in X$. Let $(Y, \Theta_Y)$ be another complex space with $y \in Y$. We say $(X, x)$ is equivalent to $(Y, y)$ if there exists some open set U containing x and a biholomorphic map between U and an open set V containing y, mapping x into y. This is an equivalent relation and an equivalence class is called a germ of a complex space. An analytic algebra is a quotient $\mathbb{C}(x_1, \ldots, x_n)/I$ where $\mathbb{C}(x_1, \ldots, x_n)$ is the ring of convergent power series in the n-variables $x_1, \ldots, x_n$ and I is a non-trivial ideal. An analytic algebra is a local $\mathbb{C}$-algebra. Let $m(A)$ be its maximal ideal, then $A/m(A) \approx \mathbb{C}$. Also, an analytic algebra is Noetherian, and is therefore separated in the Krull topology. If A and B are two local algebras and $f: A \to B$ is a homomorphism then f is local and is continuous in the Krull topology. The category of analytic algebras is dual to the category of complex spaces, as seen from the following theorem (Malgrange [562], p.20).

THEOREM 6

To any germ $(X, x)$ of a complex space there is associated an analytic ring $O_{X,x}$. Every analytic ring is obtained in this way. Every morphism $(X, x) \to (Y, y)$ of germs of complex spaces induces a homomorphism $O_{Y,y} \to O_{X,x}$ of analytic rings. Conversely, every homomorphism $B \to A$ of analytic rings is obtained from a morphism of corresponding germs of complex spaces, and the latter is unique.

## I-1-C-(3) Some further important theorems and construction theorems

1) Let $f: X \to Y$ be a holomorphic of complex spaces. The rank of f at $x \in X$, denoted by $r_x(f)$, is defined by $r_x(f) = \dim_x X - \dim_x f^{-1}(f(x))$. The rank of f, denoted by $r(f)$, is the supremum of $r_x(f)$, $x \in X$. The degeneracy set of f is defined by $E_f = \{x \in X \mid r_x(f) < r(f)\}$. It can be proved that $E_f$ is a complex analytic set. If $E_f = \phi$, f is said to be a non-degenerate map. If B is a complex analytic set in Y, it is easily proved that $f^{-1}(B)$ is a complex analytic set in X. However the image of a complex analytic set in X by f need not be a complex analytic set in Y. If in addition f is proper (that is the inverse image of any compact set by f is compact) this is true. This is a deep theorem of Remmert [782].

THEOREM 7 (Remmert's Proper Mapping theorem)

Let $f: X \to Y$ be a proper holomorphic mapping. Let A be any complex analytic set in X. Then $f(A)$ is a complex analytic set in Y whose dimension is $r(f) \mid A$. If A is irreducible, so also is $f(A)$. (The condition that f is proper is essential.)

2) Normal complex spaces: Let X be a reduced complex space. X is said to be normal if the local ring $\Theta_{X,x}$ is integrally closed in its ring of quotients. If X is normal X is locally irreducible. The set N of points x of X where X is not normal ($\Theta_{X,x}$ is not integrally closed) can be proved to be a complex analytic set. It is even analytically rare, that is the map $\Gamma(\Theta_x, U) \to \Gamma(\Theta_x, U-N)$ is injective for every open set U of X. Every reduced complex space can be normalized in the following sense.

THEOREM 8 (Normalization theorem)

For any reduced complex space X, there exist

i) a normal complex space $\tilde{X}$ (which is unique up to a biholomorphic map)

ii) a holomorphic map $f: \tilde{X} \to X$ such that f is finite and surjective and

iii) $f^{-1}(N)$ is complex analytic set which is analytically rare in $\tilde{X}$ and the restriction of f to $X - f^{-1}(N)$ gives a biholomorphic map of $\tilde{X} - f^{-1}(N)$ onto $X - N$.

If $f: X \to Y$ is a proper surjective holomorphic map between reduced irreducible complex spaces, we may seek conditions under which this is actually a biholomorphic map. The next theorem, which is a special case of a more general theorem (of Zariski) in algebraic geometry, answers this question under very restrictive conditions.

THEOREM 9 (Zariski's Main Theorem)

Let $f: X \to Y$ be a proper surjective holomorphic map between reduced irreducible complex spaces. Let (i) Y be normal (ii) f be finite (iii) there exist an open set U of Y such that $f^{-1}(u)$ is just one point for each $u \in U$. Then f is biholomorphic.

This theorem also gives information regarding when the fibres of f are connected. Before stating this we have to state a general construction theorem called the Stein factorization theorem.

THEOREM 10 (Stein factorization theorem)

Let $f: X \to Y$ be a holomorphic map of reduced complex spaces. Then there exist a reduced complex space Z, and holomorphic maps $g: X \to Z$, $h: Z \to Y$ such that

i) $f = g \circ h$

ii) g is surjective and each fibre of g is connected

iii) h is surjective and each fibre of h consists of finite number of points

iv) The number of points of $h^{-1}(y)$ is equal to the number of connected components of $f^{-1}(y)$, for each $X \to Y$.

From the Stein factorization theorem and Zariski's Main Theorem, the following theorem can be proved.

THEOREM 11 (Zariski's connectedness theorem)

Let $f: X \to Y$ be a proper surjective map of connected reduced complex spaces where Y is normal. Assume further that there exists an open set U of Y with the property for any $u \in U$, $f^{-1}(u)$ is connected. Then all the fibres of f are connected.

3) Meromorphic maps. Meromorphic maps are fundamental for the classification of compact complex manifolds. We adopt the definition of a meromorphic map due to Remmert, and state only those properties we need. For proofs and more details the reader is referred to Remmert [782], Stein [873]. The complex spaces considered in this section are all reduced and irreducible.

DEFINITION

A proper surjective holomorphic map $f: X \to Y$ between complex spaces is called a proper modification if there exist nowhere dense analytic subsets A of X and B of Y such that f is a biholomorphic map of X - A onto Y - B.

DEFINITION

Let X and Y be complex spaces. A mapping $\varphi$ from X into the power set of Y is called a meromorphic mapping of X into Y if the following two conditions are satisfied:

a) The graph $G_\varphi$ of $\varphi$ ($G_\varphi = \{(x,y)\ X \times Y \mid y \in \varphi(x)\}$ is an irreducible analytic subset of $X \times Y$;

b) The projection map $p_1 : G \to X$ is a proper modification.

A meromorphic mapping $\varphi$ of X into Y is also denoted by the usual notation $\varphi: X \to Y$. By the definition a meromorphic map $\varphi: X \to Y$ determines an irreducible analytic subset of $X \times Y$. Conversely, every irreducible analytic subset G of $X \times Y$ such that the projection (to the first factor) $p_1 : G \to X$ is a proper modification defines a meromorphic map $\varphi$ (we have only to set $\varphi(x) = p_2 \circ p_1^{-1}(x)$) whose graph $G_\varphi$ is precisely the given G.

It can be checked that a meromorphic map $\varphi: X \to Y$ is a holomorphic map if and only if $p_1$ is biholomorphic.

Note

A meromorphic function f on X is nothing but a meromorphic map $f: X \to P^1$, such that $f(X) \neq \infty$. The following theorem is important. Recall that X is said to be normal if the local ring $\Theta_{X,x}$ is integrally closed its complete ring of quotients, for every $x \in X$.

THEOREM 12 (Remmert)

Let $\varphi: X \to Y$ be a meromorphic map. Then there exists a smallest nowhere dense analytic subset $S(\varphi)$ in X such that $\varphi$ is a holomorphic map of $X - S(\varphi)$ into Y. If X and Y are both normal, then the co-dimension of $S(\varphi)$ is at least two.

One of the difficulties of meromorphic maps is that the composition of two meromorphic maps is not defined in general: if $\varphi: X \to Y$, $\Psi: Y \to Z$, $\Psi \circ \varphi$ need not be defined. However, if the projection $p_2: G \to Y$ is surjective, then $\Psi \circ \varphi$ is defined as follows:

If $x \in X' = X - (S(\varphi) \cup \varphi^{-1}(S(\Psi))$, define $\Psi \circ \varphi(x) = \Psi(\varphi(x))$.

If $x \in S(\varphi) \cup \varphi^{-1}(S(\Psi))$, define $\Psi \circ \varphi(x)$ to consist of all points $z \in Z$ such that there exist a point $y \in Y$ and a sequence $\{x_v\}$ in X such that $\{x_v\}, \{\varphi(x_v)\}$,

$\{(\Psi(\varphi(x_v))\}$ converge respectively to x,y,z (see Stein [873]. A meromorphic mapping $\varphi: X \to Y$ is bimeromorphic mapping if the projection $p_2: G \to Y$ is also proper modification. In this case we can define a meromorphic mapping $\varphi^{-1}$ from Y into X such that $\varphi . \varphi^{-1} = id_y$ and $\varphi^{-1} . \varphi = id_x$. Two complex analytic spaces are bimeromorphically equivalent if there exists a bimeromorphic mapping between them.

Meromorphic mappings arise naturally:

a) Let $f_1, \ldots, f_n$ be meromorphic functions on a compact complex space X. Then there exists an analytic set S in X outside which $f_1, \ldots, f_n$ are holomorphic. Let $X^1 = X - S$ and let G be the graph of the holomorphic mapping:

$$f : X \to \mathbb{C}^n$$

$$x \to (f_1(x), \ldots, f_n(x))$$

Let $\bar{G}$ be the closure of G in $X \times P^n(\mathbb{C})$. Then $\bar{G}$ is an irreducible analytic set in $X \times P^n(\mathbb{C})$ such that the projection $p_1 : \bar{G} \to X$ is a proper modification. Hence $\bar{G}$ determines a meromorphic map $X \to P^n_{(\mathbb{C})}$, which we denote by

$$x \to (1; f_1(x); \ldots ; f_n(x))$$

a) Let $L \to X$ be a holomorphic line bundle on a compact compex space X. Let $f_o, f_1, \ldots, f_n$ be linearly independent global sections of this line bundle. Then

$$\frac{f_1}{f_o}, \frac{f_2}{f_o}, \ldots, \frac{f_n}{f_o}$$

are meromorphic functions and hence, as explained above, these give rise to a meromorphic mapping:

$$\varphi_L ; X \to P^n(\mathbb{C})$$

$$x \to (f_o(x); f_1(x); \ldots ; f_n(x))$$

Remark

If X and Y are complete algebraic varieties, the meromorphic mappings from X into Y are exactly the rational mappings from X into Y (see e.g. Lang [519]).

4) Monoidal transforms and resolution of singularities Consider a complex manifold and a coherent sheaf g of ideals of $\mathcal{O}_X$. Let Y be the set of common zeros of

the functions of g. We can cover X by open sets U such that for each U, there exist sections $f_1, \dots f_m \in \Gamma(U,g)$ which generate $g_x$ for each $x \in U$. Define a map $f: U - U \cap Y \to P^{m-1}(\mathbb{C})$ by $f(x) = (f_1(x): \dots ; f_m(x))$. This is a holomorphic map. Let $Z'$ be the graph of f and let Z be the closure of Z in $U \times P^{m-1}(\mathbb{C})$. Z is in general a complex analytic space. Z depends only on the sheaf g and not on the particular open set U of the covering. Then Z, as U runs over the covering, can be patched up to give a complex analytic space $\tilde{X}$ and a holomorphic map $f: \tilde{X} \to X$. $\tilde{X}$ is called the generalized monoidal transform of X with respect to the sheaf g and this process of getting $\tilde{X}$ is called the generalized monoidal transformation with respect to the given g. The complex analytic set Y has its sheaf of ideals and, if g happens to be this particular sheaf itself, $\tilde{X}$ is said to be the monoidal transform of X with centre Y. If Y is nonsingular, $\tilde{X}$ is a complex manifold. If Y reduces to a single point,, $\tilde{X}$ is said to be obtained by a $\sigma$-process (or quadratic transformation) of X with respect to this point.

The monoidal transformation of a complex analytic space X with respect to a coherent sheaf g of ideals of $\Theta_X$ is similarly defined. It can be checked that in this case $\tilde{X}$ depends only on the support of the quotient sheaf $(\Theta_X/g)$ rather than on g itself. In this case we say that $\tilde{X}$ is obtained by blowing up X with centre supp $(\Theta_X/g)$.

We describe the blowing up process and its inverse process, called blowing down, in Section 2, Chapter III, in terms of local coordinates.

Hironaka's theorem on the resolution of singularities

The singularities of a complex analytic space can be resolved in the following sense.

THEOREM 13 (Hironaka)

For any complex analytic space X, there exists a proper modification $\pi: X \to X$ with X nonsingular such that over any relatively compact open set U of X, $\pi$ is the product of finite sequence of blow ups. $\pi_1 : X_i \to X_{i-1}$, $X_o = X$, with nonsingular centres $Y_{i-1}$ along which $X_{i-1}$ is normally flat.

Remark

1) This is a very deep theorem of modern algebraic geometry and it is a landmark. Hironaka's resolution theorem is considered to be a big achievement of recent times.

For the details of the proof see Hironaka [365], [366], [367], [368].

2) $X_{i-1}$ is normally flat along $Y_{i-1}$ means that $I^n_y \mid I^{n+1}_y$ is flat over $\Theta_{X_{i-1},y}/I_y$ for all points $y \in Y_{i-1}$ and for all positive integers n, where I is the sheaf of ideals defined by $Y_{i-1}$.

## I-1-C-(4) Algebraic Varieties And Projective Varieties

In the rest of this Section k will stand for an algebraically closed field of characteristic $p \geq 0$. The main references are Hartshorne [356], Mumford [641] and Shafarevich [803].

### (1) The evolution of the concepts of abstract variety and scheme

An affine algebraic set V in the affine n-space $k^n$ is the set of common zeros $(x) = (x_1, \ldots, x_n) \in k^n$ of a family of polynomials $P_\alpha$ with coefficients in k. Taking the affine algebraic sets as closed sets one can define a topology on $k^n$ which is called the Zariski topology. It is a $T_1$-topology always but in general not a $T_2$-topology (that is: it is not Hausdorff). Unless otherwise stated all the topological concepts (of varieties) that we will speak of will be with reference to this topology only.

A nonempty affine algebraic set is said to be irreducible if it is not the union of two proper algebraic subsets. An irreducible affine algebraic set is called an affine variety. An irreducible component of an affine algebraic set V is a maximal irreducible subvariety of V. Any affine algebraic set can be expressed uniquely as the irredundant union of a finite number of irreducible components. The product of two affine varieties is naturally defined and is an affine variety.

Let V be an affine algebraic set defined by a family of polynomials $f_\alpha$. The polynomials $f_\alpha$ determine the affine algebraic set V but the affine algebraic set V does not determine the polynomials $f_\alpha$ uniquely since the points of V also satisfy other polynomial equations, for example $\Sigma r_\alpha P_\alpha = 0$. To remedy this, we consider ideals I in the polynomial ring $k[x_1, \ldots x_n]$. Let $V(I) = \{(x) \in k^n \mid P(x) = 0, P \in I\}$. Then V(I) is an affine algebraic set determined by I. Let V be an affine algebraic set and let I(V) be the set of polynomial P such that $P(x) = 0$ for all $(x) \in V$. Then I(V) is an ideal in $k[x_1, \ldots, x_n]$. Thus ideals and affine algebraic sets are associated to each other. However this association is not bijective since $\mathcal{I} \neq I(V(\mathcal{I}))$

in general. But $\mathfrak{J} \subset I(V(\mathfrak{J}))$ always. More importantly, we can prove that $I(V(\mathfrak{J}))$= Radical of $\mathfrak{J}$ (= $P_\alpha \mid P_\alpha^r \in g$ for some r). This is the famous Hilbert Nullstellensatz. (Here the assumption k is algebraically closed is used, for proof see e.g. Zariski-Samuel [810]. Hence for ideals I such that I = Radical of I, the correspondence between ideals and algebraic sets is bijective. Even this is not quite a satisfactory situation since one and the same algebraic set V may have to be described by different ideals depending on the ambient space.

For example a unit circle in the complex plane is given by the ideal g generated by $x^2 + y^2 - 1$ but when considered in 3-dimensional space it is given by the ideal I generated by Z and $x^2 + y^2 - 1$. To avoid such a situation, we consider quotient rings of the form $k[x_1, \dots x_n]/\mathfrak{J}$. Now associate to each affine algebraic set V, the corresponding ring $A = k[x_1 \dots x_n]/\mathfrak{J}$ which is of finite type. If $x \in V$, the set $M_x$ of all functions in V which vanish at x is a maximal ideal in A. To an algebraic subset W of V, associate the ideal of functions vanishing on W. Conversely, given a ring of finite type $A = k[x_1 \dots x_n]/\mathfrak{J}$, we would like to associate an affine algebraic set V resulting in a bijective correspondence between affine algebraic sets and rings of finite type. Our first impulse is to take V to be the set of all maximal ideals of A. But then any non-trivial field would correspond to only a single point. This situation could be avoided by taking V to be the set of pairs $(m, A_m)$ where m is a maximal ideal in A and $A_m$ is the localization of A at m. This would be the sought after V provided distinct rings correspond to distinct sets of pairs $(m, A_m)$. Examples show that the latter possibility exists. Hence we conclude that it is no good to regard the set of pairs $(m, A_m)$ as a discrete collection. We must glue them together by a suitable topology.

Let V now denote the set of maximal ideals of A. V = Specm (A) = the set of all maximal ideals of A. For any ideal I of A, let V (I) denote the set of all maximal ideals which contain the given I. These sets V (I) define a topology on V whose closed sets are precisely these V (I). This is called the Zariski topology on Specm (A). We define a sheaf S on V such that the stalks $S_m$ at the point $m \in V$ is the local ring $A_m$. A basis of open sets in V is given by sets of the form $U(f) = \{m \mid f \notin m, f \in A\}$. Equivalently U(f) can also be described as the sets of the form $\{x \mid f(x) \neq 0\}$ where x is a 'point' of V and f(x) is the class of $f \in A$

modulo the maximal ideal x. Now the association $U(f) \to A_f$ defines a presheaf $\tilde{A}$ on V whose sections over open set $U(f)$ is $A_f$ ( $\Gamma(U(f),\tilde{A}) = A_f$). $\tilde{A}$ is in fact a sheaf. The pairs $(V,\tilde{A})$ are ringed spaces in the sense we have defined already. Since $A \approx \Gamma(V,\tilde{A})$, we have established the desired bijective correspondence between the rings A and the ringed spaces $(V,\tilde{A})$. We call $(V,\tilde{A})$ an abstract (pre) variety.

The necessity of considering abstract (pre) varieties was first recognized by A. Weil [955] and the abstract varieties that we have at in the above fashion are due to Serre. These are very important concepts of modern algebraic geometry. In Serre's definition of abstract [pre] varieties, the local rings are all reduced: they do not have any nilpotent elements. Grothendieck noted that it is necessary to allow the local rings to have nilpotent elements since they show up naturally in many geometric contexts. For example closed subvarieties of a variety naturally give rise to "varieties with nilpotent elements" (non-reduced schemes). The rings $Z/(_p n)$ for $n > O$ do have nilpotent elements. Even if two rings $A_1, A_2$ have no nilpotent elements, their tensor product $A_1 \otimes A_2$ may very well have nilpotent elements. Moreover by the introduction of nilpotent elements, instead of the situation getting more complicated, there are certain technical advantages ; especially when we deal with differentiable properties of functions.

For example if A is the ring of holomorphic functions at the origin in $C^n$ having no singularities at the origin and if $m^r$ is the maximal ideal of holomorphic functions vanishing at the origin up to order r, the local rings $A/_m r$ can be regarded as analogues in algebraic geometry of the Taylor series development in Analysis. This also allows us to speak of 'the infinitesimal neighbourhoods' (for more on the infinitesimal neighbourhoods see Section 1 of Chapter II). Among other things, with these considerations in mind, Grothendieck introduced the very important concept of a Scheme. "In the case of algebraic geometry there is no doubt that the introduction of schemes has revolutionized the subject and has made possible tremendous advances" observed Hartshorne [359] page 59. Even though a scheme is a very general concept, it was observed later by Matsusaka [582] and Artin [38], that from the point of view of the moduli problems, the notion of a scheme is not general enough. Matsusaka's Q-varieties and Artin's Algebraic spaces are further generalizations of

Schemes. It is completely outside the scope and intentions of this volume to give a systematic account of these theories. We content outselves by giving a rapid introduction to varieties and schemes. For a systematic account of algebraic varieties and schemes, two excellent references are Shafarevich [803] and Hartshorne [359]: for Q-varieties, the reference is Matsusaka [582]; and for algebraic spaces the references are Artin [38], [39] and Knutson [470].

There is also a compelling need to study algebraic geometry over non-algebraically closed fields because of natural connections to Number Theory. A. Weil [955] laid the foundations and developed a theory of algebraic geometry over arbitrary fields. Nagata [656] developed algebraic geometry over Dedekind domains. Since our main consideration is the classification problem of algebraic geometry we do not consider these generalized algebraic geometric theories.

(2) Abstract varieties

Let V be an affine algebraic variety. Its ideal I is a prime ideal. The ring $k[x_1, \dots, x_n]/I$ is called the coordinate ring of V, and it is denoted by $I[V]$. Elements of $I[V]$ can be naturally regarded as functions on V with values in k. $I[V]$ is an integral domain and its quotient field, denoted by $I[V]$, is called the (rational) functions field of V. By the Hilbert Basis theorem $I(V)$ has finitely many generators in $k[x_1, \dots, x_n]$. A subfield $k_o$ of k is said to be a field of definition for V if $I(V)$ admits generators in $k_o[x_1, \dots, x_n]$ itself. If $k_o$ is a field of definition for V, V is said to be defined over $k_o$. Define $I_{k_o}(V) = \{g \in k_o[x_1, \dots, x_n], g(x) = 0,$ for every $(x) \in V\}$ and $k_o[V] = k_o[x_1, \dots, x_n]/I_{k_o}(V)$.

$k_o(V)$ is also an integral domain. Any affine algebraic set can be shown to admit a smallest field of definition which is of finite type over the prime field. A point of V is called a $k_o$-rational point of V if all its coordinates belong to the subfield $k_o$ of k.

Let $k_o$ be a field of definition for the affine variety V. Take a point $(y) = (y_1, \dots, y_n) \in V$ and consider the homomorphism $k[x_1, \dots, x_n] \to k[y_1, \dots, y_n]$ defined by $x_i \to y_i$. This homomorphism can be factorized in a natural way:

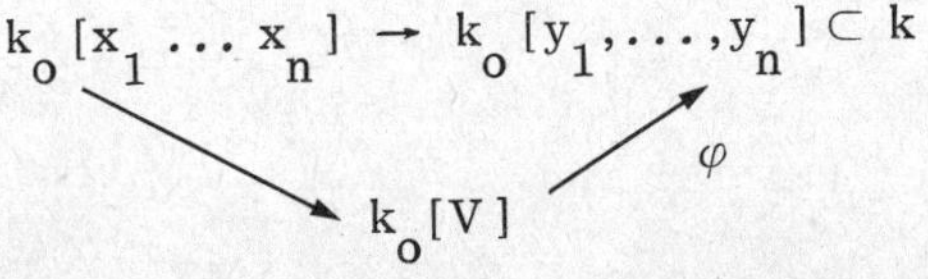

$$k_o[x_1 \dots x_n] \to k_o[y_1, \dots, y_n] \subset k$$
$$k_o[V] \xrightarrow{\varphi}$$

The point $(y)$ is said to be a generic point of V over $k_o$ if $\varphi$ is an isomorphism. A generic point need not always exist. However, if the transcendence degree of k is infinite it can be proved that for any field of definition $k_o$ for V, generic points exist for V and k.

The dimension of an affine variety V is defined to be the transcendence degree of the function field $k(v)$ (which is of finite type over k) over k. This can be proved to be equal to the Krull dimension of the coordinate ring. A variety V has dimension zero if and only if it is a point (( ===> $\mathcal{J}(V)$ is a maximal ideal). A variety V has dimension $(n-1)$ if and only if its ideal $\mathcal{J}(V)$ is principal. Such varieties are called hypersurfaces. A hypersurface can be defined by a single equation.

We now formulate the definition of an abstract variety motivated by the remarks made at the beginning of this sub-section.

DEFINITION (Serre)

An abstract prevariety X over k is a ringed space $(X, \Theta_X)$ such that there exists an open covering $(U_i)$ of X with the property that for each i, there exists an affine variety $V_i$ over k such that the restricted ring space $(U_i, \Theta_x | U_i)$ is isomorphic to the ringed space $(V_i, \Theta_{V_i})$ of the affine variety where the local ring $\Theta_{V_i} = k[V_i]_{P_i}$, $P_i$ being the ideal of functions vanishing on $V_i$.

$\Theta_X$ is called the sheaf of germs of regular functions of the variety. The concepts of irreducibility, morphisms, subvarieties, product varieties, fields of definition and generic points can be analogously defined for abstract prevarieties as in the case of affine varieties.

DEFINITION (Serre)

An abstract prevariety $(X, \Theta_X)$ is said to be an abstract variety if the diagonal map $\Delta : X \to X \times X$ given by $x \to (x,x)$ is a closed map (in the Zariski topology of $X \times X$).

This condition is analogous to the Hausdorff separation condition that we have assumed as part of the definition in the case of manifolds and complex analytic spaces.

DEFINITION

An abstract variety is said to be complete if for every abstract variety Y, the projection map $X \times Y \to Y$ is a closed map (in the Zariski topology).

This condition of being complete is analogous to the compactness condition for manifolds and complex spaces in the usual topology.

A theorem of Nagata [657] says that any abstract variety can be embedded as an open dense subset of a complete variety.

Simple and singular points: Let $(X, \Theta_X)$ be an abstract variety defined over k. A point $x \in X$ is called a simple point of X (or X is nonsingular or smooth at x) if the local ring $\Theta_{X,x}$ is a regular local ring. This means that $\dim_k ({}^{m}x/_{m_x}{}^{2}) = \dim \Theta_{X,x}$ where $m_x$ is the maximal ideal of $\Theta_{X,x}$. As in the case of complex spaces, the Zariski tangent space of X at x, denoted by $T_x X$, is defined to be the k-dual of the vector space $({}^{m}x/_{m_x}{}^{2})$. For any point $x \in X$, $\dim T_x X \geq \dim X$. If x is a simple point, then and only then $\dim T_x X = \dim X$. The set of singular points of X can be proved to be a closed subset of X strictly contained in X. A variety is nonsingular (or smooth) if all its points are nonsingular.

Biregular Maps and Birational maps: Hereafter, by variety, we would mean abstract algebraic variety: a morphism $f: (X, \Theta_X) \to (Y, \Theta_Y)$ is continuous map $f: X \to Y$ such that $f^*: \Theta_{Y,f(x)} \to \Theta_{X,x}$. A morphism is called a regular map. A regular map is said to be biregular if the induced map $f^*: \Theta_{Y,f(x)} \to \Theta_{X,x}$ is an isomorphism of local algebras for each x. Varieties and regular maps form a category and an isomorphism in this category is a biregular map.

Let $f: X \to Y$ be a regular mapping of irreducible varieties. Let $f(X) = Y$. Let $\dim X = m$, $\dim Y = n$. Then $n \leq m$. It can be further checked that (1) for every $y \in Y$, the dimension of the fibre $f^{-1}(y)$ (which is a closed subvariety of X) is at least $(m - n)$. (2) there exists a nonempty open subset U of Y such that $\dim f^{-1}(y) = m - n$ for each $y \in Y$. In general the dimension of a fibre can jump.

Birational maps: Let $f: X \to Y$ be a regular mapping. In general $f(X)$ need not be closed in Y. If X is complete this is so. The closure of $f(X)$ in Y is called the closed image of X. A closed subset Z of the product variety $X \times Y$ is called an algebraic correspondence between X and Y. If $(x,y) \in Z$, x and y are said to correspond to each other. Assume Z is irreducible and that the closed image of the projection $Z \to X$ is X itself. Then it can be proved that the function field $k(X)$

can be identified with a subfield of the function field $k(Z)$. If, moreover, $k(X) = k(Z)$, the correspondence is said to be rational mapping. If the above also holds for Y, f is said to be a birational mapping and X and Y are birational. Hence if X, Y are birational to each other their function fields are the same: $K(X) = K(Y)$. The converse is also true. Another equivalent characterization of birationality is that two varieties X, Y are birational if and only if there are open sets $U \subset X$, $V \subset Y$ such that U is biregular to V. Birational maps are also referred to as birational transformations or birational correspondences. In the case of varieties defined over C, the varieties are complex spaces and rational (birational) maps are meromorphic (bimeromorphic) maps of complex spaces which we have already discussed in the previous section I-1-C-(3).

(3) We note the following important points:

(1) A rational map is in general not a map (in the usual sense) of X to Y.

(2) A bijective rational map need not be birational.

(3) A regular map can be regarded as a special rational map by identifying it with its graph.

(4) A rational mapping of X to Y can be defined as the closure of the graph of a regular mapping of X to Y defined on an open subset of X.

(5) A rational map of X to Y can also be equivalently defined as an equivalence class of pairs $\langle U, f_U \rangle$ where U is a nonempty open subset of X, $f_U$ is a regular map from $U \to Y$ where $\langle U_1, f_{U_1} \rangle$ and $\langle U_2, f_{U_2} \rangle$ are considered equivalent if $f_{U_1}$ and $f_{U_2}$ are equal on $U_1 \cap U_2$.

(6) Rational maps cannot be composed in general.

We say that a rational map $f: X \to Y$ is dominant if for the same pair $\langle U, f_U \rangle$, the image $f(U)$ is dense in Y. Dominant rational maps can be composed in a natural way.

(7) The category of varieties defined over k and dominant rational maps is equivalent to the category of finitely generated field extensions of k.

As in the case of complex spaces, we have the notions of variety normal at a point and a normal variety. As in the case of complex spaces, every variety has a normalization.

THEOREM 14 (Zariski's Main theorem)

Let $f: X \rightarrow Y$ be a birational map. Let $x \in X$ be isolated in $f^{-1}(f(x))$ and let Y be normal at $f(x)$. Then f is biregular at x. If Y is normal any bijective birational map is biregular.

4. Rational and unirational Varieties An irreducible variety X over k whose function field $k(X)$ is purely transcendental over k is called a rational variety. If $k(X)$ has a finite algebraic extension which is purely transcendental over k, then X is said to be unirational. It is clear that every rational variety is unirational. But there are unirationals which are not rational (see e.g. Artin-Mumford [43]). A cubic three-fold is a nonsingular hypersurface in $P^4(C)$ of degree 3. Cubic three-folds are known to be unirational (see e.g. Roth [796]). It was a longstanding problem until 1972 to decide whether or not they are rational. In 1972 Clemens and Griffiths [165] proved that cubic three-folds are not rational. Their work is considered to be very important, not only because of the solution of a longstanding problem, but also because of some of the essentially new ideas introduced in proving the main result. For a detailed discussion of the fundamental work of Clemens and Griffiths, see, in addition to their original paper, Tjurin [919]. See also Murre [655], Conte-Murre [176] and Iskovoski-Manin [433] for further results in this direction.

Any unirational curve must be rational. Any unirational surface defined over an algebraically closed field of characteristic zero is known to be rational (see e.g. Shafarevich [803]). But there are examples of unirational surfaces defined over algebraically closed fields of positive characteristic which are not rational. The question whether or not a unirational variety is rational is equivalent to the Luroth problem of field theory.

5. The Problem of Resolution of Singularities. Given an irreducible variety X, the problem of resolution of the singularities of X is to find a proper birational map $f: X' \rightarrow X$ with $X'$ nonsingular. When X is a curve, it has been known for a long time that this is possible. When X is a complex analytic surface (irreducible variety of dimension 2 defined over C) with singularities, the resolution problem was solved by Walker, and later Zariski solved the resolution problem for surfaces defined over algebraically closed fields of characteristic zero. The resolution problem was solved

by Hironaka [365] for varieties of any dimension defined over algebraically closed fields of characteristic zero. Abhyankar [4] solved the resolution problem for surfaces over algebraically closed fields of positive characteristic. He also solved the problem for three-folds defined over algebraically closed fields of characteristic greater than 5. The resolution problem is one of the central and challenging problems of modern algebraic geometry. The contributions of Abhyankar and Hironaka have been tremendous. An indispensable tool in all the available proofs of the solution is the blowing up (monoidal transformations) which we have already mentioned with reference to complex manifolds. There are similar constructions for varieties. The original references are Abhyankar [2], [4], [5], Hironaka [365], [366], [367]. See also Lipman [550].

6. Divisors, Linear systems and Differential Forms. Divisors: Let X be an irreducible variety of dimension n. Consider an r-dimensional irreducible subvariety of X which is not contained in the singular locus of X. Consider the set of such subvarieties and let $D_r$ be the free abelian group generated by this set of subvarieties. Elements of $D_r$ are called the r-cycles on X. Let $A, B \in D_r$. Then we can write $A = \Sigma m_i A_i$, $B = \Sigma n_i A_i$ where $A_i$ are r-dimensional subvarieties of X not contained in the singular locus and $A_i \neq A_j$, for $i \neq j$. We write $A \geq B$ if $m_i \geq n_i$ for all i. An (n-1) dimensional cycle is called a divisor. A divisor D is said to be positive (or effective) if $D \geq 0$.

It is clear that a non-zero $\phi \in K(X)$ determines a divisor denoted by $(\phi)$. A divisor is called principal if it is the divisor of a function $\in k(X)$. Two divisors $D_1, D_2$ are said to be linearly equivalent, written as $D_1 \sim D_2$, if $D_1 - D_2 = (\phi)$ for a non-zero $\phi \in K(X)$. This is an equivalence relation. A divisor D which is linearly equal to the O-divisor in a neighbourhood of each point of X is called a Cartier divisor on X. If X is nonsingular any divisor is a Cartier divisor. Let $f: X \to Y$ be a rational map and assume that X is normal. Let D be a Cartier divisor on Y such that the closed image of f is not contained in D.

Then it can be proved that there exists a unique inverse Cartier divisor $f^{-1}(D)$ on X. The group of linear equivalence classes of Cartier divisors can be identified with the group $H^1(X, O_X^*)$, where $O_X^*$ is the sheaf of multiplicative groups of invertible elements of $\Theta_X$ (Picard group of X).

Linear systems: Let us now assume that X is a complete irreducible variety defined over k. A linear system $\mathcal{L}$ on X is a set of divisors of the form $\sum_0^n \alpha_i f_i + D$ where $\alpha_i \in k, f_0, \ldots, f_n \in k(X)$, D is a divisor on X such that $(f_i) + D \geq O$ for all i. Note that every divisor in $\mathcal{L}$ is positive and any two divisors of $\mathcal{L}$ are linearly equivalent. $\mathcal{L}$ is said to be complete if any positive divisor D on X which is linearly equivalent to a divisor in $\mathcal{L}$ belongs to $\mathcal{L}$. Any linear system can be uniquely extended to a complete linear system $|\mathcal{L}|$. A divisor D is called a fixed component of $\mathcal{L}$ (or $|\mathcal{L}|$) if for any divisor $E \in \mathcal{L}$, $E > D$. Let $D_o$ be the maximal fixed component with respect to the order $\geq$. $D_o$ is called the fixed component of $\mathcal{L}$. For any $D \in \mathcal{L}$, $D\text{-}D_o$ is called the variable component of $\mathcal{L}$ (or D).

A point $x \in X$ is called a base point of a linear system if x is on every variable component of $\mathcal{L}$. $\mathcal{L}$ is said to be irreducible if it has no fixed component and if its generic member is irreducible. The dimension of the linear space $\sum_{i=O}^n k_{f_i}$ (which is called the defining module of $\mathcal{L}$) be denoted by $\ell(\mathcal{L})$. Then the dimension of is defined to be $\ell(\mathcal{L})-1$. For a divisor D, $|D| = \{E, \text{ a divisor } | E > O \text{ and } E \sim D\}$ is called the complete linear system defined by D.

Ample divisors: Let $f_o, \ldots, f_n$ be a basis for a defining module of $\mathcal{L}$. Then we get a rational map $\Phi(\mathcal{L}) : X \to \mathbb{P}^n(k)$ given by $x \to (f_o(x); \ldots, f_n(x))$. $\Phi(\mathcal{L})$ is a morphism (regular) in the complement of the base locus of $\mathcal{L}$. If $\Phi(\mathcal{L})$ is biregular then $\mathcal{L}$ is is said to be very ample. For a divisor D, if $|D|$ is very ample we say that D itself is a very ample divisor. If for some integer m, mD is very ample, we say that D is ample.

Differential forms: Let X be an n-dimensional algebraic variety defined over k. The set of derivations of the function field k(X) is an n-dimensional linear space Der(X) and let its dual be denoted by $\mathrm{Der}^*(X)$. Recall a derivation of k(X) is a k-linear map $\lambda: k(X) \to k(X)$ such that $\lambda(f,g) = \lambda(f)g + f\lambda(g)$. For each $f \in k(X)$ define $df \in \mathrm{Der}(X)$ by $\langle df, \lambda \rangle = \lambda(f), \lambda \in \mathrm{Der}^*(X)$. Let $x_1, \ldots, x_n$ be algebraically independent over k and the function field k(X) be a separable algebraic extension of $k(x_1, \ldots, x_n)$. Such $x_1, \ldots, x_n$ exist and are called a separating transcendence basis. Then $dx_1, \ldots, dx_n$ forms a basis for Der(X) over k(X). Consider the Grassmann algebra $\mathcal{G}$ (Der(X)) of Der(X) over k(X). The homogeneous elements of degree r of $\mathcal{G}$ (Der(X)) are called differential forms of degree r on X, and is denoted by $\Omega^r[X]$.

Let $f_1, \ldots, f_n \in k(X)$ be such that on an open set U of X, $f_1 - f_1(x), \ldots, f_n - f_n(x)$ is a regular system of parameters for the local ring $D_x$ for each $x \in U$. If such $(f_1, \ldots, f_n)$ exists it is said to be a system of local uniforming coordinates. It can be proved that (1) A system of local uniformizing coordinates is a separating transcendence basis of k(X); (2) For any simple point $x \in X$, there exists a system of local uniformizing coordinates in a neighbourhood of x; (3) Any differential form of degree r on X can be written (as in the case of manifolds) as

$$w = \sum_{i_1 < \ldots < i_r} \varphi_{(i)} \, df_{i_1} \wedge \ldots \wedge df_{i_r} .$$

If the coefficients $\varphi_{(i)}$ are regular functions at x, the form w is said to be regular at x. The set of points at which a regular differential form vanishes is closed, since it corresponds precisely to the set of points x such that $\varphi_{(i)} = 0$.

Given a differential form w on a complete irreducible variety X, one can associate in a canonical way a divisor on X, denoted by $|w|$. The divisor of a differential form of top degree (= n) is called a canonical divisor. The canonical divisors of degree n form a linear equivalence class of divisors.

In case X is complete and non-singular, this can be seen easily as follows. Locally $w = g\, df_1 \wedge \ldots \wedge df_n$ be a differential form of degree n. Cover X with affine open sets U such that in each open set U the above representation holds. On $U_1 \cap U_j$,

$$g^{(j)} = g^{(i)} \quad J\left(\frac{f_1^{(i)}, \ldots, f_n^{(i)}}{f_1^{(j)}, \ldots, f_n^{(j)}}\right),$$

where the Jacobian is non-zero and regular. The system of functions $g_1^{(i)}, \ldots, g_n^{(i)}$ defines the divisor (w).

Consider a regular mapping $f: X \to Y$ of varieties. This induces in a natural way a map $f^*: \Omega^r[Y] \to \Omega^r[X]$. In general nothing can be said about this mapping. If, however, the function field $k(X)$ has a separating transcendence basis over $k(Y)$, then $f^*$ is an embedding. $\Omega^r[X]$ can be shown to be a birational invariant.

Let $w_1, w_2$ be two differential forms on open sets $U_1, U_2$, respectively. We say $(w_1, U_1) \sim (w_2, U_2)$ if $w_1 = w_2$ on $U_1 \cap U_2$. This is an equivalence relation and an equivalence class is called a <u>rational</u> form. The space of all rational differential forms on $X$ is denoted by $\Omega^r(X)$. $\Omega^r(X)$ is a linear space of dimension $\binom{n}{r}$ over $k(X)$.

<u>7. Serre Duality Theorem</u> Analogous to the situation over complex numbers, on an irreducible algebraic variety defined over an algebraically closed field K, we have the notions of algebraic sheaves, algebraic vector bundles, coherent algebraic sheaves, cohomology with coefficients in a sheaf of germs of sections of algebraic vector bundles (refer to Chapter III of Hartshorne [359]).

An invertible sheaf on X is a coherent sheaf of a $\Theta_X$-module which is locally free of rank 1. One can prove that on a non-singular X, the three notions of divisors, invertible sheaf and algebraic line bundles are all equivalent. A line bundle L (locally free sheaf of rank 1) is ample if the associated divisor, denoted by $|L|$, is ample. For a line bundle L the associated locally free sheaf is denoted by $\underline{L}$.

For a complete irreducible algebraic variety $(X, \Theta_X)$, the cohomology groups $H^q(X, \Theta_X)$ are well defined and they are all finite dimensional. Let $h^{o,q}$ = $\dim H^q(X, \Theta_X)$. The integer $\sum_{q=0}^{n} (-1)^q h^{o,q}$ is called the Euler-Poincare characteristic of X and is denoted by $\chi(X)$. It is also called the arithmetic genus of X and is denoted by $p_a(X)$. It is a birational invariant of X. Sometimes $(-1)^n(\chi(X)-1)$ is also defined to be $p_a(X)$.

Let X be a non-singular irreducible complete variety and let $\underline{\Omega^p(X)}$ denote the sheaf of germs of regular differential forms of degree p. Then $\dim H^q(X, \underline{\Omega^p(X)})$ is finite and is denoted by $h^{p,q}(X)$.

<u>THEOREM 15 (Serre Duality)</u>

Let X be a non-singular complete irreducible variety of dimension n, A be any algebraic vector bundle over X and $A^*$ be the dual vector bundle of A. Let $\underline{A}$ $(\underline{A}^*)$ denote the corresponding sheaf of germs of sections of A (respectively $A^*$). Then the vector space $H^q(X, \underline{A})$ is canonically isomorphic with the vector space

$H^{n-q}(X, \underline{A}^* \otimes \underline{\Omega}^p(X))$.

Note that $H^n(X, \underline{\Omega^n(X)}) \approx k$. It follows from the above theorem that $H^q(X, \underline{\Omega^p(X)})$ is dual to $H^{n-q}(X, \underline{\Omega^{n-p}(X)})$. Hence $h^{p,q}(X) = h^{n-p,n-q}(X)$. If $k = C$, X is a compact connected complex manifold and further assume that X is Kahler, then one can prove $h^{p,q}(X) = h^{q,p}(X)$ by Hodge theory. In particular this is true for algebraic surfaces defined over an algebraically closed field of characteristic zero. However if k is of positive characteristic this symmetry breaks down (e.g. Serre [833]).

8. Projective Varieties. The main reference for this section is Mumford [641]. K continues to stand for an algebraically closed field.

The projective n-space $P^n(K)$ over K is the quotient of $K^{n+1} - (o)$ for the action by multiplication of $K^*$, the group of non-zero elements of K. Any point $(x) \in P^n(K)$ can be represented (up to multiplication by elements of K) by an (n+1) -tuple $(x_o, \ldots, x_n)$ $x_i \in K$, with at least one $x_i \neq O$. $(x_o, \ldots, x_n)$ are called the homogeneous coordinates. A projective algebraic set in $P^n(K)$ is defined to be the set of points $(x) \in P^n(K)$ such that $F_\alpha(x) = O$ for a family of homogeneous polynomials $(F_\alpha)$, $F_\alpha \in K[x_o, \ldots, x_n]$. The Zariski topology on $P^n(K)$ is the topology whose closed sets are precisely the projective algebraic sets in $P^n(K)$. The ideal generated by all homogenous polynomials in $K[x_o, \ldots, x_n]$ vanishing on V is called the homogeneous ideal of V and is denoted by $\mathcal{H}I(V)$. The ring $\mathcal{H}I[V] = K[x_o, \ldots, x_n]/\mathcal{H}I(V)$ is called the homogeneous coordinate ring of V. I(V) has a natural grading determined by the grading in the polynomial ring $K[x_o, \ldots, x_n]$. I(V) is a graded ring. As in the affine case we have the correspondence between ideals in $K[x_o, \ldots, x_n]$ and projective algebraic sets in $P^n(K)$ with an important difference: in the affine case, an ideal I determines the empty set if and only if $I = (1)$. In the projective case, a homogeneous ideal I determines the empty set if and only if it contains the ideal $I_r \subset K[x_o, \ldots, x_n]$ consisting of those homogeneous polynomials all of whose terms are of degree $\geq r$ $(> O)$.

An irreducible projective algebraic set in $P^n(K)$ is called a projective variety. If V is a projective variety, $\mathcal{H}I(V)$ is seen to be a prime ideal not containing the ideal $(X_o, \ldots, X_n)$ and the converse is also true.

For studying the connections between affine varieties and projective varieties, we note a basic important fact: every projective variety V has an open (Zariski) covering by affine varieties. To see this first note that there exists a homeomorphism of $k^n$ onto the open set $X_o \neq O$ in $P^n(K)$ given by $(y_1, \ldots, y_n) \rightarrow (1, y_1, \ldots, y_n)$. Let the image be $k_o$. The hyperplane $H_o$ given by $X_o = O$ is called the hyperplane at infinity for the affine subspace $K_o$. Similarly we have hyperplanes $H_i = X_i = O$ whose complements $K_i$ are affine. Then $P^n(K) = \bigcup_{n=O}^{n} K_i$. Hence any projective variety V in $P^n(K)$ can be expressed as $\bigcup_{i=O}^{n} V_i$ where $V_i = V - H_i = V \cap K_i$. Each $V_i$ is an affine algebraic subset of $K_i$. Using this important connection, we can define for a projective variety V, local rings of V, the function field of V, tangent spaces to V, the dimension of V, singular points of V, normalization of V and the various results relating to these remain valid with suitable modification. In particular normalization of a projective variety exists and is a projective variety.

We can also define in a natural way products of projective varieties: the product of projective varieties is a projective variety. We can also define correspondences and rational maps. In particular we have projective versions of the Zariski Main theorem and the Zariski connectedness theorem. The results on divisors, linear systems and differential forms also remain valid with the necessary modifications.

The following important points should be noted:

(1) A projective variety V is a variety in the sense defined in the previous subsection, for it can be proved that the diagonal maps $\Delta: V \rightarrow V \times V$ is closed.

(2) The image of a projective variety under a regular mapping is closed.

(3) Any projective variety is complete in the sense defined earlier: for any variety W, the projection map $V \times W \rightarrow W$ is a closed map. If the field K is the field of complex numbers (or real numbers), completeness is equivalent to compactness in the usual topology.

(4) Any nonsingular projective variety of dimension n can be biregularly embedded in $P^{2n+1}(K)$.

(5) It can be proved that two nonsingular projective varieties defined over C are biholomorphic (as complex manifolds) if and only if there exists a biregular correspondence between them as projective varieties.

## 9. Finiteness theorems of Serre and Grothendieck.

THEOREM 16 (Finiteness theorem of Serre)

Let X be a projective algebraic variety over an algebraically closed field K and $\mathcal{S}$ be an algebraic coherent $\Theta_X$-module. Then $H^q(X, \mathcal{S}) = O$ for $q > \dim X$ and $H^q(X, \mathcal{S})$ are finite dimensional K-vector spaces for $O \leq q \leq \dim X$.

Grothendieck generalized this theorem to any algebraic variety. Let $f: X \to Y$ be a proper morphism of algebraic varieties over K. If  is any $\Theta_X$-module, the sheaf associated to the presheaf $U \to H^p(f^{-1}(U), \mathcal{S})$ is called the $p^{th}$ direct image sheaf of  and is denoted by $R^p f_*(\mathcal{S})$ (The induced map $f_*$ is a left exact additive functor from $\Theta_X$-modules to $\Theta_Y$-modules. Later on in the subsection on schemes we will see $R^p f_*$ is the $p^{th}$ derived function of $f_*$).

THEOREM 17 (Finiteness Theorem (Directo Image Theorem) of Grothendieck)

Let $f: X \to Y$ be a proper morphism of algebraic varieties over K. $\mathcal{S}$ be an alge raic coherent $\Theta_X$-module. Then its direct images $R^p f_*(\mathcal{S})$ are all coherent $\Theta_Y$-modules.

Taking Y to be a point, we get a finiteness theorem for complete varieties X. If X is projective the above reduces to Serre's Finitenesstheorem. Later we shall state a further generalization of the above theorem to preschemes.

10. Criteria for projectiveness. It is an important problem to decide when an arbitrary complete variety is a projective variety. We give two criteria: first applicable to nonsingular complete varieties and the second to complete varieties (singularities may be present).

THEOREM 18 (Chevalley-Kleiman Criterion)

A nonsingular complete irreducible variety X is projective if and only if for any set S of points of X, there exists an affine open set $S'$ containing S.

There is also a version of the criterion when V has singularities. See Kleiman [465].

To state our next criterion we have to introduce the concept of a positive line bundle as defined by Grauert. Let $L \to X$ be an algebraic bundle over a complete irreducible variety X. Let $L^* \to X$ be the dual line bundle. Let Z be the zero section of

$L^* \to X$: by definition Z is the section which associates to each $x \in X$, the zero of the fibre $L^*_x$ at x. We say that Z can be blown down to a point if there exists a regular map $\varphi: L^* \to Y$ where Y is the affine variety containing the origin in some $K^m$ such that $\varphi(z) = O$ and $\varphi|_{L^*-Z}: L^* - Z \to Y - (O)$ is a biregular isomorphism. If Z can be blown down to a point we say that $L^* \to X$ is negative and $L \to X$ is a positive line bundle. We have the following important criterion of Grauert [293].

Grauert's criterion for projectiveness: A complete irreducible variety is projective if and only if there exists a positive line bundle on it.

Remark

In the case of a complete irreducible variety X (nonsingular or otherwise) defined over C (i.e., in the case of a compact complex manifold or space), this question of deciding when X is projective is of paramount importance. We take up this question in Section 3.

11. Quasi-projective varieties. Affine algebraic varieties and projective varieties can be simultaneously studied by studying quasi-projective varieties. A quasi-projective variety is defined to be an open subset of a closed projective set. If a quasi-projective variety X is isomorphic to a closed subset of an affine space, then X is an affine variety. If X is isomorphic to a closed projective set, then X is a projective variety. However there are quasi-projective varieties which are neither affine varieties nor projective varieties (e.g. $K^2 - (x)$, $P^2(K) - (x)$ where x is a point). There are also abstract varieties which are not isomorphic to quasi-projective varieties.

12. The classification problem in Algebraic Geometry. "In any branch of mathematics, there are usually guiding problems, which are so difficult that one never expects to solve them completely, yet which provide stimulus for a great amount of work and which serve as yardsticks for measuring progress in the field. In algebraic geometry such a problem is the classification problem". R. Hartshorne [356], p. 55.

The classification problem is the problem of classifying all algebraic varieties up to biregular equivalence. A less ambitious problem is the problem of classifying all algebraic varieties up to birational equivalence and then classifying in each birational class the non-singular projective varieties up to projective equivalence (isomorphism). In spite of the tremendous efforts of outstanding mathematicians, 'complete' success has been achieved so far only in dimensions 1 and 2. A brief account of this success story is given in Chapter III with particular reference to varieties defined over C. In order to tackle the general classification problem, a general procedure is to look for invariants (discrete and/or continuous) for varieties with which non-isomorphic varieties could be distinguished. Another important goal is to give a structure of algebraic variety (and then study its properties) to the set of all distinct isomorphism classes of varieties. This is the so-called Moduli Problem. An account of the success achieved so far, relating to this problem, is given in Section 4 of this Chapter.

## SCHEMES, PROJECTIVE SCHEMES AND ALGEBRAIC SPACES

I-1-C-(5) Throughout this subsection A will stand for a commutative ring with unit 1. All rings considered are commutative rings with unity.

(1) Pre-schemes and schemes. The set of prime ideals ($\neq 1$) of A is denoted by Spec A. Let X = Spec A. Since A has at least one maximal ideal, and therefore at least one prime ideal, $X \neq \emptyset$. For any subset E of A, let $V(E) = \{x \in X \mid \text{the prime ideal } x \supseteq E\}$. Taking the sets V(E) as closed sets a topology on X can be defined and this topology is called the Zariski (or spectral) topology on X. We consider only this topology on X.

For $f \in A$, let f(x) denote the class of f module the prime ideal x in A/x. Then f(x) = O and only if f belongs to the ideal x. Let the complement of the closed set V(f) in X be denoted by D(f). The open sets $D(f) = \{x \in X \mid f(x) \neq O\}$ form a basis for the (Zariski) topology of X. It is easy to check that if $x, y \in X$, then $y \in \{\overline{x}\}$ if and only if (the prime ideal) $x \subseteq$ (the prime ideal) y. Hence $\{x\}$ is a closed set if and only if the ideal x is maximal. If $\{x\}$ is a closed set, x is called a closed point of X. Hence X is a $T_1$ space (a topological space in which every point is a closed set) if and only if every ideal of A is maximal.

We now define a sheaf of rings on X. Let $f, g \in A$ such that $D(f) \supseteq D(g)$. Then $r(f) \geq r(g)$ where r(f) is the radical of f consisting of those elements of A such that some power of it equals f (and a similar interpretation for r(g)). Hence there exist $s \in A$ and $n > 0$ such that $g^n = sf$. Define a ring homomorphism $\rho_{g,f}$ between the ring of fractions $A_f$ and $A_g$ as:

$$\rho_{g,f} : A_f \to A_g$$

by setting $f_{g,f}(a/_f m) = a\, s^m /_g mn$. This homomorphism is well defined and depends only on f and g. It is also easy to check that if $D(f) \supseteq D(g) \supseteq D(h)$, then $\rho_{h,g} \circ \rho_{g,f} = \rho_{h,f}$. Hence the assignment $D(f) \to A_f$, together with the homomorphisms $f_{g,f}$ defines a presheaf on X which is actually a sheaf $\Theta_X$. The sections over an open set $D(f)$ denoted by $\Gamma(D(f), \Theta_X)$, is $\approx A_f$ and $\Gamma(X, \Theta_X) = A$. Thus we have now got a ringed space $(X, \Theta_X)$ starting from a ring A. The stalk $\Theta_x$ at any point $x \in X$ is the local ring $A_x$ (localization of A at the prime ideal x). It is easy to check that if $\varphi: A \to B$ is a homomorphism of rings, then $\varphi$ induces a homomorphism of the ringed spaces $(\text{Spec } B, \Theta_{\text{Spec } B}) \to (\text{Spec } A, \Theta_{\text{Spec } A})$ and any homomorphism of the later ringed spaces is induced by a homomorphism of the rings $A \to B$.

Hereafter, by the spectrum of A, we will mean the ringed space $(\text{Spec } A, \Theta_{\text{Spec } A})$.

DEFINITION

(1) An affine scheme is a locally ringed space $(X, \Theta_X)$ which is isomorphic to the spectrum of some ring A.

(2) A pre-scheme is a locally ringed space $(X, \Theta_X)$ such that for each $x \in X$, there exists an open set U with the property that $(U, \Theta_{X|U})$ is an affine scheme.

A morphism of pre-schemes is a morphism of locally ringed spaces. Pre-schemes and morphisms of pre-schemes form a category.

Let S be a fixed pre-scheme. By a pre-scheme over S we mean a pre-scheme X together with a morphism $X \to S$. If X, Y are two pre-schemes over S, an S-morphism between X and Y is a morphism $X \to Y$ compatible with the given morphism: that is

$$\begin{array}{ccc} X & \rightarrow & Y \\ & \searrow \; \nearrow & \\ & S & \end{array}$$

is commutative.

Pre-schemes over S and morphisms over S form a category. The following theorem connects varieties and pre-schemes. Let Pre-var (K) be the category of prevarieties over K , (algebraically closed) and Pre-sch (K) be the category of preschemes over K. Then:

THEOREM 1

There is a natural (fully-faithful) functor $\tau$; pre-var (K) $\to$ pre-sch (K) such that if V is a pre-variety over K then its underlying topological space is homeomorphic to the closed points of the topological space X underlying the pre-scheme $\tau(V) = (X, \Theta_X)$ and the sheaf $\Theta_V$ of regular functions on V is obtained by restricting the structure sheaf $\Theta_X$ of the pre-scheme $\tau(V)$ via this homeomorphism.

Let X, Y be two S-pre-schemes. Then one can prove that the fibre product $X \times_S Y$ exists and is unique up to a canonical isomorphism. Let $S_1, S_2$ be two pre-schemes and $S_1 \to S_2$ be a morphism. Let $X_2$ be an $S_2$-pre-scheme. Let $X_1 = X \times_{S_2} S_1$. Then $X_1$ is a $S_1$-pre-scheme. $X_1$ is said to be got from $X_2$ by making a base extension $S_1 \to S_2$.

The product of two schemes X and Y is defined to be the fibre product $X \times_{\mathrm{Spec}\, Z} Y$.

Let $f : X \to S$ be a morphism of pre-schemes. Consider the diagonal morphism $\Delta : X \to X \times_S X$. If $\Delta(X)$ is closed in $X \times_S X$, we say f is separated. A pre-scheme X is said to be a scheme if the characteristic morphism $X \to \mathrm{Spec}\ Z$ is separated. (Each pre-scheme admits a unique morphism to Spec Z called the characteristic morphism). As in the case of varieties, this condition is the analogue of the Hausdorff condition. Any affine scheme is separated.

Now let $(X, \Theta_X)$ be any scheme. Then an open subscheme of X is a scheme $(Y, \Theta_Y)$ with underlying topological space Y being an open subset of X, and the structure sheaf $\Theta_Y$ being isomorphic to $\Theta_{X|U}$. An open immersion is a morphism $f : Y \to X$ which induces an isomorphism of X with an open subscheme of X.

A closed subscheme of a scheme $(X, \Theta_X)$ is a scheme $(Z, \Theta_Z)$ such that the underlying topological space Z of $(Z, \Theta_Z)$ is a closed subset of the underlying topological space X of $(X, \Theta_X)$ and the induced sheaf map $i^* : \Theta_X \to i_* \Theta_Z$ is surjective, i being the inclusion map $Z \to X$. A closed immersion is a morphism $f : Z \to X$ which induces an isomorphism of Y onto a closed subscheme of X. It is to be noted from this definition that any closed subset Z of X has many closed subscheme structures.

A subscheme W of a scheme X is an open subscheme of a closed subscheme of X. If W is a subscheme of X, the induced map : $W \to X$ is called an immersion.

We say a scheme X is :

(1) Connected if its topological space is connected.

(2) Irreducible if its topological space is irreducible.

(3) Reduced if the local rings $\Theta_{X,x}$, $x \in X$ have no nilpotent elements. (As in the case of complex spaces associated to any scheme there is a reduced scheme $X_{red}$).

(4) Integral if it is both reduced and irreducible.

(5) Noetherian if X can be covered by a finite number of affine subsets space $A_i$ with each $A_i$ Noetherian.

(6) A morphism $f: X \to Y$ of schemes is said to be of finite type if Y can be covered by open affine subsets $V_i$ = Spec $B_i$ such that $f^{-1}(V_i)$ is affine and equal to Spec $A_i$, say, and $A_i$ is a $B_i$-algebra and is a finitely generated $B_i$-module.

(7) A morphism $f: X \to Y$ of schemes is said to be proper if (i) it is of finite type; (ii) it is closed and (iii) for any morphism $Y' \to Y$, the corresponding morphism obtained by base extension is also closed. If conditions (ii) and (iii) together hold we say that if f is universally closed. Hence a morphism $f: X \to Y$ of schemes is proper if it is of finite type and universally closed.

(8) In view of the importance of proper morphisms, it is useful to have a good criterion for properness. The most useful criterion is :

THEOREM 21 (Valuation Criterion for properness)

Let $f: X \to Y$ be a morphism of finite type with X Noetherian. Then f is proper if and only if for every valuation ring R with quotient field K and for every commutative diagram

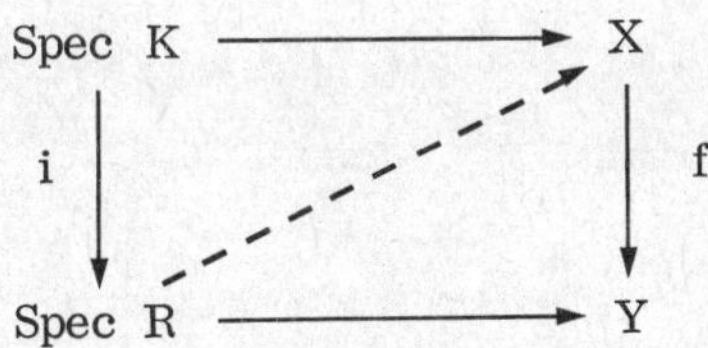


there exists a unique morphism Spec R $\to$ X making the entire diagram commutative.

Using this criterion we can show the following important properties : composition of proper morphisms is proper, products of proper morphisms is proper and proper morphisms are stable under base change.

(9) A scheme of finite types over Spec K, where K is an algebraically closed field, is called an algebraic scheme over K or an algebraic K-scheme. A proper algebraic

K-scheme is called a complete scheme.

THEOREM 22

The category of reduced algebraic K-schemes and the category of algebraic varieties defined over K are equivalent.

(10) A morphism $f : X \to Y$ of schemes is said to be flat if for each point $x \in X$, $\Theta_{X,x}$ is flat $\Theta_{Y,f(x)}$-module. If further f is surjective, f is said to be faithfully flat. As in the case of flat holomorphic maps, flat morphisms of schemes have analogously nice properties, see Chapter II, section 1 of this volume, and Hartshorne [356], Chapter III, Section 9.

(11) Let $(X, \Theta_X)$ be a scheme. Let $k(x)$ be the residue class field $\Theta_{X,x}$. For $f \in \Theta_{X,x}$, the residue class of f in $k(x)$ is called the value of f at x, denoted by $f(x)$. Note Spec $k(x)$ is a single point ; there exists a natural morphism i ; Spec $k(x) \to X$ whose image is the singleton set $(x)$. Let K be an algebraically closed field. A morphism $i : \mathrm{Spec}\ K \to X$ is called a geometric point of X. A geometric point is determined by a point x of X and an embedding $k(x)$ in K. Let $i : \mathrm{Spec}\ K \to S$ be a geometric point of S. A geometric fibre for i is $X \times_S \mathrm{Spec}\ K$.

(12) Recall that the dimension of a topological space X is the supremum of all positive integers n such that there exists a chain $X_o \subset X_1 \subset \ldots \subset X_n$ of distinct irreducible closed subsets of X. The dimension of a scheme $(X, \Theta_X)$ is defined to be the dimension of the underlying topological space X. If X is an affine scheme Spec A, the dimension of X is the same as the Krull dimension of A. If X is an integral scheme of finite type, over a field K, one can show that at any closed point $x \in X$, the Krull dimension of the ring $\Theta_{X,x}$ is equal to the dimension of X.

The Zariski tangent space of $(X, \Theta_X)$ at $x \in X$, denoted by $T_x X$, is defined to be the dual of the $k(x)$-vector space $m_x / m_x^2$. If $f : X \to Y$ is a morphism of schemes, there is a natural induced mapping $T_f : T_x X \to Y_y Y \otimes_k k(x)$.

Analogous to varieties, on a schème we have the notions of divisors, linear systems, locally invariable sheaves, vector bundles, ample sheaves (divisors) and ample invertible sheaves. In particular there exist the divisor class groups on a scheme X : there exists a one-to-one correspondence between isomorphism classes of locally free sheaves of rank n on a scheme X and isomorphism classes of vector bundles of rank n on X.

(2) Projective Schemes

Let $A = \bigoplus_{d \geq O} A_d$ be any graded ring. Let $A_+ = \bigoplus_{d > O} A_\alpha$ be the ideal of positive elements of $\bar{A}$. Let Proj A denote the set of all homogeneous prime ideals which do not strictly contain $A_+$. It is easy to see that Proj A = $\phi$ if and only if every element of $A_+$ is nilpotent. For any homogeneous ideal a of A set $V(a) = \{p \in \text{Proj } A \mid p \supseteq a\}$. Taking $V(a)$ as the closed subsets, we can define a topology on Proj A. We can define a sheaf of rings on Proj A such that the stalk $\Theta_p$ at $p \in$ Proj A is the local ring $A_p$ (localization of A at p). The local ringed space (Proj A, $\Theta_{\text{Proj } A}$) has in fact the structure of a scheme. If A is a ring, then the polynomial ring $A[x_o, \ldots, X_n]$ is a graded ring and the associated scheme Proj $A[x_o, \ldots X_n]$ is called the projective n-space of A and is denoted by $P^n_A$. This is consistent with our earlier definition of a projective space over an algebraically closed field in the sense that the subspace of all closed points of $P^n_K$ is homeomorphic to the projective n-space $P^n_K$ as defined earlier. More generally, if V is a projective variety over K with homogeneous coordinate ring A, then the associated scheme $\tau(V)$ is (isomorphic to) Proj A. For any ring A, we can show that $P^n_A \approx P^n_Z \times$ Spec A. For any scheme X, the projective n-space over X, denoted by $P^n_X$; is defined to be $P^n_Z \times X$. A morphism $f: X \to Y$ of schemes is said to be projective if for some n it can be factored as

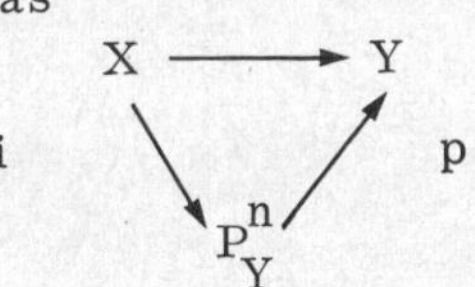

where i is a closed immersion and p is the projection.

A morphism $f: X \to Y$ is said to be quasi-projective if it can be factored as

$X \longrightarrow Y$, $j: X \to X$, $q$

where j is an open immersion of schemes and q is a projective morphism. The following are important properties of projective morphisms and quasi-projective morphisms.

THEOREM 23

Any projective morphism of Noetherian pre-schemes is proper. Any quasi -

projective morphism of Noetherian pre-schemes is separated and is of finite type.

Theorem 20 on the functor $\tau$ can be strengthened as follows.

THEOREM 24

Let K be an algebraically closed field. Then the image of the functor $\tau$: Pre-var (K) $\to$ Pre-sch (K) is the set of all quasi-projective integral schemes over K. The image of the set of all projective varieties over K is the set of all projective integral schemes.

From the above theorem it follows that the scheme associated to an irreducible variety is an integral scheme of finite type over K. However, an abstract irreducible variety over K (in the sense defined earlier) is an integral scheme of finite type over K.

THEOREM 25

A scheme X over Spec A is projective if and only if there exists a graded ring B such that $X \approx \text{Proj } B$ with $B_o = A$ and B is finitely generated as a $B_o$-module by $B_1$.

The twisting sheaf $\Theta(1)$. Let $A = \bigoplus_{d>O} A_d$ be any graded ring and M be a graded A-module. M defines a unique sheaf $\mathcal{M}$ on X = Proj A as follows. For any $p \in \text{Proj } A$, let S be the multiplicative systemof homogeneous elements not in p. Let $\mathcal{M}_p$ be the group of elements of degree 0 in the localization of $S^{-1}(M)$. For an open set $U \subseteq X$, let $\mathcal{M}(U)$ be the set of functions $f : U \to \amalg_{p \in U} \mathcal{M}_p$ which are locally fractions. The association $U \to \mathcal{M}(U)$ defines a presheaf and it is in fact a sheaf $\mathcal{M}$ of $\Theta_X$-module. In particular take $M = A_n$. Then the associated sheaf on X is denoted by $\Theta_X(n)$. If A is generated by $A_1$ as an $A_o$-module, then $\Theta_X(n)$ are all invertible sheaves (locally free sheaves of rank 1) and $\Theta_X(n) \otimes \Theta_X(m) = \Theta_X(n+m)$. The sheaves $\Theta_X(1)$ are particularly important in the study of projective schemes. For any sheaf of $\Theta_X$-modules F, let $F(n) = F \otimes_{\Theta_X} \Theta_X(n)$. For any scheme Y, consider the projective space $P^r_Y = P^r_Z \times_{\text{Spec } Z} Y$. We have a natural map $f : P^r_Y \to P^r_Z$. The pull back sheaf $f^*(\Theta(1))$ on $P^r_Y$ of the sheaf $\Theta(1)$ on $P^r_Z$ is also denoted by $\Theta(1)$. Now let X be any scheme over Y. An invertible sheaf L on X is said to be very amply relative to Y if for some r there exists an immersion $i : X \to IP^r_Y$ such that

$i^*(\Theta(1) \approx L$. In particular the sheaf $\Theta(1)$ on $X = \text{Proj } A$ and $X = P^n_K$, where K is a field, is very ample. $\Theta(n)$ are also ample for $n > O$.

THEOREM 26

Let A be a ring and X be a scheme over A. Then if L is an invertible sheaf on X such that there are a finite number of global sections $s_o, \ldots, s_n$ of L generating L, there exists a unique A-morphism $\varphi: X \to \mathbb{P}^n_A$ such that $L = \varphi^*(\Theta(1))$ and moreover, $s_i = \varphi^*(x_i)$ under this isomorphism where $x_i$ are the homogeneous co-ordinates of $\mathbb{P}^n_A$ and, conversely, any A-morphism $\varphi: X \to P^n_A$ is such that the sheaf $\varphi^*(\Theta(1))$ is an invertible sheaf generated by the global section $\varphi^*(x_i)$.

THEOREM 27

Let X be any scheme over a Noetherian scheme Y. Then X is projective if and only if it is proper and there exists a very ample sheaf on X relative to Y.

An invertible sheaf L on a Noetherian scheme X is said to be ample if for any co-herent sheaf F on X, there exists an integer $n_o > 0$ such that for all $n \geq n_o$, $F \otimes L^n$ is generated by its global sections.

THEOREM 28 (Serre)

A very ample sheaf on a projective scheme X over a Noetherian ring A is ample. The converse of Theorem 28 is false in general.

THEOREM 29

Let X be a scheme of finite type over a Noetherian ring A and L be an invertible sheaf on X. Then L is ample if and only if $L^m = L \otimes \ldots \otimes L$ (m times) is very ample over Spec A for some $m > O$.

(3) Quasi-coherent and coherent sheaves

As in the case of complex spaces and varieties there exists the important notion of a coherent sheaf over a scheme. First we shall define a more general notion of a quasi-coherent sheaf.

Let $(X, \Theta_X)$ be a scheme and F be a sheaf of $\Theta_X$-modules over X. A $\Theta_X$-module homomorphism $\alpha: \Theta_X \to F$ gives rise to a global section s of F and one can show that every global section of F comes from a $\Theta_X$-module homomorphism $\alpha: \Theta_X \to$ . We have only to set $\alpha(t) = t(s|U)$ for an open set U of X. Hence there exists a one-to-one correspondence between $\Theta_X$-module homomorphisms $\alpha: \Theta_X \to F$ and global sections of $F$. Hence there exists a one-to-one correspondence between $\Theta_X$-module homomorphisms $\alpha: \Theta_X^I \to F$ and families $(s_i)_{i \in I}$ of global sections of $F$, where I is any indexing set. $\alpha$ will be surjective if and only if $F_x$ is generated as an $\Theta_X$-module by $(s_i)_x$ for each $x \in X$.

F is said to be quasi-coherent for each $x \in X$, if there exists a neighbourhood U such that $F \mid U$ is the co-kernel of a homomorphism $\Theta_X^I \mid_U \to \Theta_X^J \mid_U$ where I and J are abitrary indexing sets depending only on U. Note $\Theta_X$ is itself quasi-coherent. The sheaves $\Theta_X(n)$ are quasi-coherent.

F is said to be of finite type if locally it is generated by a finite set of local sections: each $x \in X$ has a neighbourhood U such that $F \mid U$ is generated by a finite number of sections of F over U.

F is said to be coherent if (i) F is of finite type and (ii) for each open set U of X, for each homomorphism $\phi: \Theta_X^n \mid U \to F \mid U$, $\operatorname{Ker} \phi$ is of finite type.

From the definition it is clear that a coherent sheaf is quasi-coherent. As in the case of analytic coherent sheaves over complex manifolds, we have the following important theorem.

THEOREM 30

Let $(X, \Theta_X)$ be a scheme. Let $O \to F \to G \to H \to O$ be an exact sequence of $\Theta_X$-modules on X. If any two of F, G, H are coherent, so is the third.

From this theorem we can deduce that if $\varphi: F \to G$ is a homomorphism of coherent $\Theta_X$-modules on X, then the kernel of $\phi$, the image of $\phi$ and the co-kernel of $\phi$ are all coherent. Also $F \otimes_{\Theta_X} G$ and $\operatorname{Hom}_{\Theta_X}(F, G)$ are coherent. We note that $\Theta_X$ itself need not be coherent. If it is, we say it is a coherent sheaf of rings.

Let X = Spec A be an affine scheme. Let F be a quasi-coherent $\Theta_X$-module on X. Then F is generated by its global sections $\Gamma(X, F)$.

THEOREM 31

There is an equivalence between the category of quasi-coherent sheaves on the affine scheme X = Spec A and the category of A-modules (given by $F \to \Gamma(X, F)$). If A is Noetherian, the coherent sheaves on X and the finite A-modules correspond to each other in this equivalence.

(4) Grothendieck cohomology.

Given any topological space X and a sheaf A of abelian groups on X, the Cech cohomology groups $H^q_{Ce}(X,A)$ are defined with respect to a covering.

Let $(X,\Theta_X)$ be a scheme. Given any $\Theta_X$-module F, there is the usual Czech cohomology group $H^q_{Ce}(X,F)$ of X with coefficients in F with respect to a covering. A disadvantage of the Czech cohomology is that the cohomology sequences arising out of an exact sequence of $\Theta_X$-modules need not be exact. Grothendieck introduced another cohomology where this defect is not present.

Let $(X,\Theta_X)$ be a ringed space. Let $\underline{C}$ be the category of $\Theta_X$-modules. It is an abelian category. An object I in $\underline{C}$ is said to be injective if the functor $A \to \text{Hom}(A,I)$ is exact. (That is whenever $A \to B$ is an injective homomorphism, the induced map $\text{Hom}(B,I) \to \text{Hom}(A,I)$ is a surjective homomorphism.) The category $\underline{C}$ has enough injectives (that is every object in $\underline{C}$ can be embedded in an injective object). Since $\underline{C}$ has enough injectives, for any object A there exists an injective resolution: a long exact sequence

$$O \to A \to I^0 \xrightarrow{\alpha^o} I^1 \to I^2 \to \ldots$$

with each $I^{(r)}$ being injective.

Let F be a covariant additive left exact functor on $\underline{C}$ with values in another abelian category. We get a complex

$$O \to F(A) \to F(I^0) \xrightarrow{F(\alpha^o)} F(I^1) \xrightarrow{F(\alpha^1)} F(I^2) \to \ldots$$

Define for $p \geq 0$, $H^p = \text{Ker}(F(\alpha^p))/\text{Im}(F(\alpha^{p-1}))$. Set $\alpha^{-1} = 0$. It can be checked that $H^p$ depends only on A and F. $H^p$ is usually denoted by $R^pF(A)$ to indicate this dependence. Given F, we get an additive functor $R^pF : A \to R^pF(A)$. $R^pF$ is called the pth right derived functor of F. Note that $R^0F = F$ and if A itself is injective then $R^pF(A) = O$ for all $p > 0$.

We have the following important theorem concerning the derived functors (for proof see Godement [281]).

THEOREM 32

Let $O \to A_1 \xrightarrow{\alpha_1} A_2 \xrightarrow{\alpha_2} A_3 \to O$ be an exact sequence in $\underline{C}$. Let F be a covariant additive left exact functor on $\underline{C}$ with values in another abelian category $\underline{C}'$.

Then there exists an exact sequence in $\underline{C}'$.

$$0 \to F(A_1) \xrightarrow{F(\alpha_1)} F(A_2) \xrightarrow{F(\alpha_2)} F(A_3) \xrightarrow{\partial} R^1F(A_1) \xrightarrow{R^1F(\alpha_1)} R^1F(A_2)$$

$$\xrightarrow{R^1F(\alpha_2)} R^1F(A_3) \xrightarrow{\partial} R^2F(A_1) \to \ldots$$

Apply the above to the section functor $\Gamma$ which associates to any sheaf S on X, its global sections $\Gamma(X, S)$. $\Gamma$ is a covariant additive left exact functor on $\underline{C}$ with values in the abelian category of $\Gamma(X,\dot{F})$-modules. The Grothendieck cohomology groups of X with coefficients F, denoted by $H^p_{Gr}(X,F)$, are defined to be the derived functors of the section functor $\Gamma$.

$$H^p_{Gr}(X,F) = R^p\,\Gamma(F),\ p \geq 0.$$

The Grothendieck cohomology groups have the following important property (which is not true in general for the Cech cohomology groups $H^p_{ce}(X,F)$.

THEOREM 33

If $0 \to F_1 \to F_2 \to F_3 \to 0$ is an exact sequence of $\Theta_X$-modules, then the following induced long sequence of cohomologies is exact

$$0 \to \Gamma(X, F_1) \to \Gamma(X,F_2) \to \Gamma(X, F_3) \to H^1_{Gr}(X, F_1) \to$$

$$H^1_{Gr}(X,F_2) \to H^1_{Gr}(X, F_3) \to H^2_{Gr}(X,F_1) \to \ldots$$

Under certain circumstances we can assert that both the Cech and Grothendieck cohomologies are the same. For example

THEOREM 34

Let $U = (U_i)$ be an open covering of X and F be any sheaf of $\Theta_X$-module such that (i) U is closed under finite intersection;

(ii) $H^q_{Ce}(U_i, F(U_i)) = 0$ for all $U_i \in U$ and $q > 0$. Then $H^q_{Ce}(X,F) \approx H^q_{Gr}(X,F)$ for all $q \geq 0$.

Let $(X,\Theta_X)$ be any scheme. As a ringed space its Grothendieck cohomology groups are denoted by $H^p_{Gr}(X, F)$. Let $(X,\Theta_X)$ be a Noetherian scheme. For each quasi-coherent $\Theta_X$-module F, the Grothendieck cohomology $H^q_{Gr}(X,F)$ can be shown to be canonically isomorphic to the Cech cohomology $H^q_{Ce}(X,F)$ defined with respect to an open affine covering. (See p. 222, Hartshorne [356]). If X is of dimension d, then $H^q(X,F) = 0$ for all $q > d$ and for any sheaf of abelian groups F on X. The maximum of q such that $H^q(X,F) \neq 0$ is called the cohomological dimension of X. The cohomological dimension of X cannot exceed the dimension of X. If X is affine, the cohomological dimension of X is zero and the converse is also true if X is Noetherian (see the following theorem of Serre). In the other extreme it can be proved that for al algebraic scheme X, the cohomological dimension of X is equal to the dimension of X if and only if X is complete. The following theorem gives a cohomological characterization of affine schemes.

THEOREM 35 (Serre)

Let X be a Noetherian scheme. Then the following conditions are equivalent.

(a) X is affine.

(b) $H^i(X,F) = 0$ for any quasi-coherent sheaf on X, $i > 0$.

(c) $H^1(X,F) = 0$ for any coherent sheaf of ideals .

We will see later in Section 3 of this chapter that there are analogous theorems in complex spaces characterizing Stein spaces.

The finiteness theorem of Grothendieck for algebraic varieties gets generalized as follows.

THEOREM 36 (Finiteness theorem or direct image theorem of Grothendieck for pre-schemes).

Let $f: X \to Y$ be a proper morphism of pre-schemes where Y is Noetherian. Let F be any coherent $\Theta_X$-module. Then its direct images $R^p f_*(F)$ are coherent $\Theta_Y$-modules.

The corresponding theorem for complex spaces was first proved by Grauert. For Grauert's theorem and its importance see Chapter II, Section (4).

THEOREM 37 (Finiteness theorem of Serre)

Let X be any projective scheme over a Noetherian ring A, and be any coherent sheaf on X. Then

(a) $H^i(X, F)$ are all finitely generated A-modules for each $i \geq 0$.

(b) Let $F(n) = F \otimes_{\Theta_X} \Theta(n)$ where $\Theta(n)$ are the local invertible sheaves on X (over Spec A) introduced earlier. Then there exists an integer $n_o$, depending on such that for each $i \geq 0$ and for all $n \geq n_o$, $H^i(X, F(n)) = 0$.

Another important theorem related to the above is the following:

THEOREM 38 (The upper semicontinuity theorem of Grothendieck)

Let $f: X \to Y$ be a projective morphism of Noetherian schemes and let F be a coherent sheaf on X flat over Y. Then for each $i \geq 0$, the functions $\dim_{k(y)} H^i(X_y, F_y)$ is an upper semicontinuous function on Y, where $X_y$ denotes the fibre at y and $F_y = F_{|X_y}$.

The upper semicontinuity theorem was first proved by Grauert for complex spaces. (See Section (4), Chapter II for the statement and its importance in deformation theory.

As in the case of varieties, singularities can be defined and the problem of the resolution of singularities can be posed for schemes defined over an algebraically closed field. Where the characteristic of the field is zero, Hironaka's theorem on the resolution of singularities holds in the case of schemes also. The process of 'blowing up' again plays a crucial role.

## (5) Smooth and etale maps

Let $f: X \to Y$ be morphisms f of schemes. Consider the diagonal morphisms $\Delta: X \to X \times X$. Then $\Delta(X)$ is a closed subscheme of an open set W in $X \times_Y X$. Let I be the sheaf of ideals of $\Delta(X)$ in W. The sheaf of relative differentials of X over Y is defined to be the sheaf $\Delta^*(I / I^2)$ on X and is denoted by $\Omega_{X|Y}$. Since $\Delta$ gives an isomorphism of X to $\Delta(X)$ and $(I / I^2)$ has a natural structure of the $\Theta_{\Delta(X)}$ module, $\Omega_{X/Y}$ acquires a natural $\Theta_X$-module structure. It can be proved that $\Omega_{X/Y}$ is quasi-coherent and if Y is Noetherian and f is of finite type then $\Omega_{X/Y}$ is coherent. If X is the affine n-space $A^n_Y$ over Y, $\Omega_{X/Y}$ becomes a free $\Theta_X$-module of rank n generated by the 'differentials' $dx_1, \ldots, dx_n$ where $x_1, \ldots, x_n$ are the affine coordinates of Y.

Smooth morphisms. A morphism $f: X \to Y$ of schemes of finite type over any field k is said to be smooth of relative dimension n if the following three conditions are satisfied:

(i) f is flat;

(ii) If $X'$, $Y'$ are irreducible components of X and Y respectively such that $f(X')=Y'$, then $\dim X' = \dim Y' + n$;

(iii) for every point $x \in X$, $\dim_{k(x)} (\Omega_{X/Y} \otimes k(x)) = n$.

For example if Y = Spec K, and K is algebraically closed, then one can check that $f: X \to Y$ is smooth is equivalent to the condition that X be regular of dimension n. Hence if X is an irreducible variety then X smooth is equivalent to the condition that X is non-singular. More precisely, we have:

THEOREM 39

Let $f: X \to Y$ be a morphism of non-singular varieties over K (algebraically closed). Let $n = \dim X = \dim Y$, then the following are equivalent (i) f is smooth of relative dimension n; (ii) $\Omega_{X/Y}$ is a locally free sheaf of rank n; (iii) for every closed point $x \in X$, the map $T_f: T_x(X) \to T_{f(x)}(Y)$ is surjective.

Smooth morphisms remain smooth under base change, composition and fibre product.

Etale maps. An etale map is a smooth map $f: X \to Y$ of relative dimension zero.

THEOREM 40.

Let $f: X \to Y$ be a morphism of finite type over an algebraically closed field K.

(A) Suppose f is etale. Then the following conditions are equivalent (i) f is flat; (ii) f is flat and $\Omega_{X/Y} = 0$ and (iii) f is flat and unramified (unramified means for each $x \in X$, $\mathfrak{m}_y \mathcal{O}_{X,x} = \mathfrak{m}_x$ and $k(x)$ is a separable algebraic extension of $k(y)$ where $y = f(x)$).

(B) f is etale if and only if for each $x \in X$, the local rings $\mathcal{O}_{X,x}$, $\mathcal{O}_{Y,y}$, $y = f(x)$, become isomorphic when completed. (This means: let $\hat{\mathcal{O}}_{X,x}$, $\hat{\mathcal{O}}_{Y,y}$ be their completions: there exists a natural map $\hat{\mathcal{O}}_{Y,y} \to \hat{\mathcal{O}}_{X,x}$. We require that $k(x)$ be a separable algebraic extension of $k(y)$ and

$$\hat{\mathcal{O}}_{Y,y} \otimes_{k(y)} k(x) \approx \hat{\mathcal{O}}_{X,x}).$$

Schemes and etale maps form an important category. There is an important cohomology theory called Etale cohomology theory with which cohomology for varieties over any field can be defined. Specifically the notion of an I -adic cohomology of a variety can be defined, which plays a crucial role in some important aspects of recent developments of algebraic geometry (e.g. Weil conjectures, classification of surfaces in characteristic $p > 0$).

(6) Formal Schemes.

We have already mentioned how a closed subvariety Y of a variety X gives rise naturally to some non-reduced schemes (called the infinitesimal neighbourhood of Y) which give information on the imbedding of Y in X. By considering the formal completion of Y in X, precise information about all infinitesimal neighbourhoods of Y in X can be got. In fact the formal completion of Y in X can be regarded as the limit of the infinitesimal neighbourhoods of Y.

Let A be a ring and I be an ideal of A. Let $\widehat{A}$ be the I-adic completion of A. $\widehat{A}$ is the projective limit of $(A/_{I}n)$ as $n \to \infty$. Taking $(I^n)$ as a fundamental system of neighbourhoods, a topology on A is defined which is called the I-adic topology. The completion of A in the I-adic topology can be shown to be isomorphic to $\widehat{A}$ (regarding $(A/_{I}n)$ as discrete topological rings). There exists a continuous homomorphism $i : A \to \widehat{A}$. If i is an isomorphism, A is said to be complete with respect to I. $\widehat{A}$ is complete with respect to $\widehat{I} = i(I)\widehat{A}$. If A is Noetherian we can prove that $\widehat{A}$ is also Noetherian.

Let $(X, \Theta_X)$ be a Noetherian scheme and Y be a closed subscheme. Let I be the ideal defining Y. The formal completion of Y along X is a local ringed space $(\widehat{X}, \Theta_{\widehat{X}})$ as follows: set $\widehat{X} = Y$; the sheaf $(\Theta_{X/I^n})$ can be considered as a sheaf of rings on Y. The sheaves $(\Theta_{X/I^n})$ form an inverse system and their projective limit is denoted by $\Theta_{\widehat{X}}$. The structure of $\Theta_{\widehat{X}}$ depends only on the closed subset Y and not on its scheme structure. The stalks of $\Theta_{\widehat{X}}$ are local rings. Let $\mathcal{F}$ be any coherent sheaf on X. The completion $\widehat{\mathcal{F}}$ of $\mathcal{F}$ along Y is the projective limit of the sheaves $(\mathcal{F} / _{I^n \mathcal{F}})$ on Y. $\widehat{\mathcal{F}}$ has a natural $\Theta_{\widehat{X}}$-module structure. We call $(\widehat{X}, \Theta_{\widehat{X}})$ a formal local model.

A Noetherian formal pre-scheme is a local ringed space $(X, \Theta_X)$ such that there exists an open covering $U_i$ of X with the property that for each i, $(U_i, \Theta_{X \mid U_i})$ is isomorphic as a local ringed space to a formal model $(\widehat{X}_i, \Theta_{\widehat{X}_i})$. A morphism between two Noetherian formal pre-schemes is defined to be a morphism of the category of local ringed spaces. Noetherian formal pre-schemes and morphisms form a category. This category strictly includes the category of Noetherian pre-schemes and morphisms. A Noetherian formal pre-scheme is called a Noetherian formal scheme if it is separated (the diagonal morphism $\Delta_X: X \to X \times X$ is closed).

Formal algebraic geometry (theory of formal schemes) has been developed analogous to the theory of schemes. We refer the readers to Hironaka-Matsumara [370] for more on this technical topic.

(7) Algebraic spaces.

An algebraic space X (in the sense of M. Artin) is a pair (U,R) where U is an affine scheme and R is a closed subscheme of $U \times U$ such that (i) R is an equivalence relation and (ii) the two projection morphisms $R \to U$ are etale. An algebraic space X is usually denoted as $R \rightrightarrows U \to X$. (For the precise interpretation of the double arrows see Artin [38] or Knudson [470]). A morphism of an affine scheme V to an algebraic space $R \rightrightarrows U \to X$ consists of a closed subscheme $W \subset U \times V$ such that (i) the projection morphism $W \to V$ is etale and (ii) the two closed subschemes $R \times_u W$, $W \times_v W$ of $U \times U \times V$ are equal. Let $S \rightrightarrows U \to Y$ be another algebraic space. Then a morphism : $Y \to X$ between these algebraic spaces is an element of the kernel of Hom (V,X) $\rightrightarrows$ Hom (S,X). Algebraic spaces and morphisms of algebraic spaces form a category (strictly adhering to an earlier definition, an algebraic space should be called a pre-algebraic space since it need not be separated). An important theorem is that the category of algebraic spaces contains the category of schemes (as a faithful and full sub-category). Most of the results of pre-schemes can be generalized to algebraic spaces by a general theory called the descent theory. An algebraic space is 'not very far' from schemes in the sense that every algebraic space has a dense open subspace which is a scheme. One of the motivations for introducing an algebraic space is to remedy the defect that when an algebraic group acts on an algebraic scheme 'the geometric quotient space' may not exist as a scheme. In the extended category of algebraic spaces such quotients exist under reasonable assumptions. See Popp [747], [749], Mumford [633] and Seshadri [840], [841], [843] and Haboush [341].

Algebraic spaces defined over C are particularly important. It can be shown that underlying any separated algebraic space X of finite type defined over C, there exists a complex analytic space $X^{an}$. An important result of Artin is that if there exists a proper modification $f: X^{an} \to Y$, where Y is also a complex analytic space, then Y is algebraisable in the sense that Y carries the structure of an algebraic space (whose underlying complex analytic space is the given Y) and f extends to a morphism of algebraic spaces. Just as there are reasons to consider formal schemes, so there are reasons for introducing the notion of a formal algebraic space. For the theory of formal algebraic spaces see Knutson [470], Chapter Five.

(8) Comparison Theorems (GAGA)

Throughout this subsection X, Y, . . . denote schemes of finite type over C. We have seen complex varieties can be regarded in a natural way as irreducible reduced schemes of finite type over C. We now associate to each scheme X of finite type over C a complex space $X_{an}$ in a natural way as follows : Let $(Y_i)$ be an affine open covering of X and let $Y_i = \operatorname{spec} A_i$ where $A_i$ are of the form

$$C[X_{i,1}, \ldots, X_{i,n_i}] / (f_{i,1}, \ldots, f_{i,q_i}).$$

The set of common zeros of $f_{i,1}, \ldots, f_{i,q_i}$ is a closed complex space $Y_{i,an}$ of $C^{n_i}$. These $Y_{i,an}$ can be patched up since $Y_i$ patch up to give the scheme X. Patching up $Y_{i,an}$ we get a complex space which does not depend on the affine covering of X but depends only on X. Denote this complex space by $X_{an}$. It is clear from the construction that complex spaces associated to isomorphic schemes are biholomorphic. It also follows that given a morphism $f: X \to Y$, there is the induced holomorphic map $f_{an}: X_{an} \to Y_{an}$. Similarly to the above construction, we can associate to each coherent sheaf S on X, a coherent analytic sheaf $S^{an}$ on $X_{an}$. Then for each i, there are maps $\alpha_i : H^i(X, S) \to H^i(X_{an}, S_{an})$. The relationship between X and $X_{an}$, S and $S_{an}$ and properties of the maps $\alpha_i$ were extensively investigated by Serre in his famous paper GAGA [831]. The following facts are easily checked.

(a) X is smooth, normal, reduced, proper and irreducible if and only if $X_{an}$ is, respectively, a manifold, normal, reduced, compact and connected.

(b) Let $f: X \to Y$ be a morphism. Then f is flat, unramified, etale, smooth, proper, open immersion, closed immersion and isomorphism if and only if $f_{an} : X_{an} \to Y_{an}$ has respectively the same property.

The following questions can be posed immediately:

1. Should every complex space be the associated complex space of a scheme of finite type over C ?
2. If the associated complex spacees of two schemes are biholomorphic, should the schemes be isomorphic ?
3. Let $\mathcal{F}$ be a coherent analytic sheaf on $X_{an}$. Should there exist a coherent sheaf S on X such that $S_{an}$ is isomorphic with $\mathcal{F}$ ?
4. If the associated coherent analytic sheaves of two given coherent sheaves on a scheme are isomorphic, should the given coherent sheaves be isomorphic ?
5. Are the maps $\alpha_i : H^i(X, S) \to H^i(X_{an}, S_{an})$ isomorphisms?

In general the answers to the above questions are in the negative. For counter examples see Safarevich [803], chapter VIII, and Hartshorne [356], Appendix B. However, in the case of projective algebraic schemes, Serre has proved that the answers to the above questions are in the affirmative. As in the case of a scheme, we can associate to each separated algebraic space X of finite type C, a complex space and properties of this association can be studied analogously. For details see Knutson [470].

We summarise by stating that the entire discussion in this Part C concerns the following structures and sequence of implications:

Complex Manifolds → Complex Spaces → Varieties|k →

Schemes|k → Algebraic Spaces|k.

APPENDIX (Section 1)

## A NOTE ON CHARACTERISTIC CLASSES

Let $\mathbb{E} : E \to X$ be a bundle over a space X with structure group G. Let $B_G$ be the classifying space for G and $f : X \to B$ be the classifying map ( which is unique up to homotopy). Then f induces a unique homomorphism $f^* : H^*(B_G, R) \to H^*(X, R)$. The elements in the image of $f^*$ are called characteristic classes of $\mathbb{E}$. The characteristic classes of a manifold are defined to be the characteristic classes of its tangent bundle. For the study of topology of manifolds, especially, characteristic classes are important. We give a brief account of Stiefel-Whitney classes, Euler class, Chern classes and Pontrjagin classes. Standard references are Kobayashi-Nomizu [475] Vol. 2 and Milnor [602].

### I. Stiefel-Whitney Classes

We define Stiefel-Whitney classes axiomatically. For existence and uniqueness see Milnor [602].

Let $\mathbb{E} : E \to X$ be a vector bundle. Consider $H^*(X, \mathbb{Z}_2)$ - the cohomology ring of X with coefficients in $\mathbb{Z}_2$ ( Integers modulo 2). Stiefel-Whitney classes of $\mathbb{E} : E \to X$ is a sequence of cohomology classes $\{ \omega_i(\mathbb{E}) \mid \omega_i(\mathbb{E}) \in H^i(X, \mathbb{Z}_2) \}$ satisfying

(i) $\omega_0(\mathbb{E}) = 1 \in H^0(X, \mathbb{Z}_2)$

(ii) $\omega_i(\mathbb{E}) = 0$, $i > n =$ rank of $\mathbb{E}$

(iii) if $\mathbb{E} : E \to X$, $\eta : F \to Y$ are two vector bundles with $f : X \to Y$ a map covered by a bundle map $\mathbb{E} \to \eta$ ; then $\omega_i(\mathbb{E}) = f^* \omega_i(\eta)$. (Naturality)

(iv) if $\mathbb{E} : E \to X$, $\eta : F \to Y$ are two vector bundles over X, then $\omega_i(\mathbb{E} \oplus \eta) = \sum_{k=0}^{i} \omega_k(\mathbb{E}) . \omega_{i-k}(\mathbb{E})$. ( The product appearing on the right hand side is the cup product in $H^*(X, \mathbb{Z}_2)$.) (Whitney Product Theorem).

Consider the projective n-space $\mathbb{P}^n(R)$. Consider the canonical line bundle $\nu'_n$ defined as follows: consider $\mathbb{P}^n(R) \times \mathbb{R}^{n+1}$. Let $\nu'_n = \{ (\bar{x}, y) \mid y \in \bar{x} \}$ :$(\nu'_n)_{\bar{x}} =$ the fibre of $\nu'_n$ at $\bar{x} = \bar{x}$ ( considered as a one dimensional vector space). Then it is required that

(v) $\omega_1(\nu'_1) \neq 0$ for $\nu'_1 \to \mathbb{P}^1(R)$. (Normalisation).

The following properties are easily checked.

(1) If $\mathbb{E} \simeq \eta$ then $\omega_i(\mathbb{E}) = \omega_i(\eta)$.

(2) If $\mathbb{E}$ is trivial, $\omega_i(\mathbb{E}) = 0$ for $i > 0$.

(3) If $\epsilon$ denotes the trivial bundle then $\omega_i(\epsilon + \eta) = \omega_i(\eta)$.

(4) If $\mathbb{E}$ admits k linearly independent sections, then

$\omega_{n-k+1}(\mathbb{E}) = \ldots = \omega_n(\mathbb{E}) = 0$.

Consider $H(X) = \prod_i H^i(X, \mathbb{Z}_2)$ (direct product). If $x = a_o + a_1 + \ldots$ and $y = b_o + b_1 + \ldots$ then $xy = a_o b_o + (a_o b_1 + a_1 b_o) + (\quad) + (\quad) + \ldots$ defines a product in $H(X)$ induced by the cup product in $H^*(X)$. As we work modulo 2, this product is commutative. The total Stiefel-Whitney class of $\mathbb{E}$ is $\omega(\mathbb{E}) = \omega_o(\mathbb{E}) + \omega_1(\mathbb{E}) + \ldots H(X, \mathbb{Z}_2)$. Then the Whitney's product theorem is $\omega(\mathbb{E} \oplus \eta) = \omega(\mathbb{E}).\omega(\eta)$.

Note that $x = 1 + x_1 + \ldots \in H(X)$ is invertible in $H(X)$. Hence if $\mathbb{E} \oplus \eta = \epsilon$ (trivial bundle) then $\omega(\mathbb{E}) = )\ \omega(\eta))^{-1}$.

Using the axioms of Stiefel-Whitney classes, we indicate the computations of Stiefel-Whitney classes of some standard bundles. If M is a manifold then by $\omega(M)$ we mean $\omega(T_M)$. The total Stiefel-Whitney classes for $S^n = \omega(S^n) = 1$. If $\nu'_n$ denotes the canonical line bundle over $\mathbb{P}^n(R)$ then $\omega(\nu'_n) = 1 + a$ where $a \in H^1(\mathbb{P}^n(R), \mathbb{Z}_2)$ is a generator. (Note that $H^i(\mathbb{P}^n(R), \mathbb{Z}_2)$ is cyclic for $0 \leq i \leq n$ and $H^i(\mathbb{P}^n(R), \mathbb{Z}_2) = 0$ for $i > n$. If $a \in H^1(\mathbb{P}^n(R), \mathbb{Z}_2)$ is a generator then $a^i$ generates $H^i$.) If $\nu^{\perp}$ denotes the orthogonal complement of $\nu'_n$ in $\epsilon^{n+1}$, the trivial bundle then $\omega(\nu^{\perp}) = (1 + a)^{-1} = 1 + a + a^2 + \ldots + a^n$. If $\tau$ denotes the tangent bundle of $\mathbb{P}^n$, then $\omega(\tau) = (1 + a)^{n+1}$. In particular we have the following result due to Stiefel: $\omega(\mathbb{P}^n) = 1$ if and only if $n + 2 = 2^t$, t a positive integer. Thus the only $\mathbb{P}^i(R)$ which may be parallelizable are those $\mathbb{P}^i(R)$ for $i = 1, 3, 7, 15, \ldots$. Note that $\mathbb{P}^1(R)$, $\mathbb{P}^3(R)$ and $\mathbb{P}^7(R)$ are actually parallelizable whereas $\mathbb{P}^{15}(R)$, $\mathbb{P}^{31}(R)$ ... are not. (Adams, Bott-Milnor, Kervaire.)

Let M denote an n-dimensional connected compact manifold, $\omega_i \in H^i(M, \mathbb{Z}_2)$, the ith Stiefel-Whitney class of its tangent bundle $\bar{\omega}_i \in H^i(M, \mathbb{Z}_2)$, the ith dual Stiefel-Whitney class. $\omega_i$ and $\bar{\omega}_i$ are related by $(\bar{z}, \omega_i)(\bar{z}, \bar{\omega}_i) = 1$.

Note that $\bar{\omega}$ can be thought of as Stiefel-Whitney class of the normal bundle of any differential embedding of M in an Euclidean space. The following properties of Stiefel-Whitney classes of M can be proved easily.

(i) if n is odd, $\omega_n = 0$.

(ii) $\omega_1 = 0$ if and only if M is orientable.

(iii) $\bar{\omega}_n = 0$ for every n.

Further properties of Stiefel-Whitney classes are given by the following theorems. ( For proofs see Massey [573] ).

THEOREM 1

Let M be a compact n-manifold and q be an integer, $0 < q < n$. If $\omega_{n-q} \neq 0$, then there exist integers $h_1, \ldots, h_q$ such that $h_1 \geq h_2 \geq \ldots \geq h_q \geq 0$ and $n = 2^{h_1} + \ldots + 2^{h_q}$.

If M is orientable the following additional restrictions must be imposed. (a) $q \neq 1$, (b) If $n \equiv 2 \bmod 4$ then $h_q \neq 1$, (c) An odd number of the $h_i$'s are not equal to $h_{q+1}$.

COROLLARY

(i) If $\bar{\omega}_{n-1} \neq 0$, then n is a power of 2 and M is nonorientable.

(ii) If $\bar{\omega}_{n-2} \neq 0$, then $n = 2^k(2^k + 1)$ for non-negative integers h and k. If M is orientable, in addition, the case $n = 2(2^k + 1)$ for $h > 0$ and $n = 3.2^k$ are not possible.

(iii) If $n = 2^r - 1$, then $\bar{\omega}_i = 0$ for $i > n-r$.

THEOREM 2

If n is even and M is orientable then $\omega_{n+1} = 0$.

THEOREM 3

If $n \equiv 3 \bmod 4$ and M is orientable then $\omega_n = \omega_{n-1} = \omega_{n-2} = 0$.

Stiefel-Whitney numbers

Let M be a closed smooth n-dimensional manifold. Let $\mu_M \in H_n(M, \mathbb{Z}_2)$ be the fundamental homology class. (For the definition of fundamental homology see Milnor [602] ). Then for every $\alpha \in H^n(M, \mathbb{Z}_2)$, $(\alpha, \mu_M)$ is in $\mathbb{Z}_2$. It is denoted by $\alpha[M]$ and is called the Kronecker index of $\alpha$.

Let $(r_1, \ldots, r_n)$ ne an n-tuple of non-negative integers such that $\sum_{i=1}^{n} ir_i = n$. Consider the monomial $\omega_1(\mathbb{E})^{r_1} \ldots \omega_n(\mathbb{E})^{r_n} \in H^n$.

DEFINITION

$\langle \omega_1^{r_1} \ldots \omega_n^{r_n}, \mu_M \rangle$ is called the Stiefel-Whitney number associated with $\omega_1^{r_1} \ldots \omega_n^{r_n}$. It is denoted also by $\omega_1^{r_1} \ldots \omega_n^{r_n}[M]$.

THEOREM 4 ( Pontrjagin-Thom)

Let M be a compact smooth n-dimensional manifold without boundary. Then $M = \partial B$, where B is another compact manifold of dimension n+1, if and only if all the Stiefel-Whitney numbers vanish.

DEFINITION

Two smooth closed n-dimensional manifolds $M_1$ and $M_2$ belong to the same oriented cobordism class if and only if their disjoint union $M_1 \cup M_2$ is the boundary of a smooth compact n=1 dimensional manifold.

From Theorem 4 we get:

THEOREM 5 (Thom)

Two closed n-dimensional manifolds belong to the same cobordism class if and only if their corresponding Stiefel-Whitney numbers are the same.

## II. Euler Class

Let V be a vector space of dimension n. An orientation in V amounts to prescribing a generator $\mu_V$ of $H^n(V, V_o, \mathbb{Z})$, $(V_o = V \setminus \{0\})$, which in turn fixes a generator $u_V \in H^n(V, V_o, \mathbb{Z})$ such that $\langle \mu_V, u_V \rangle = +1$.

THEOREM 6 ( Thom Isomorphism Theorem)

Let $\mathbb{E}: E \to B$ be an oriented n-plane bundle. Then (i) $H^i(E, E_o, \mathbb{Z}) = 0$, $(E_o = E \setminus \{0\})$, for $i < n$. (ii) $H^n(E, E_o, \mathbb{Z})$ contains a unique cohomology class u such that $u|_{(F, F_o)} \in H^n(F, F_o, \mathbb{Z})$ is the prescribed generator. (iii) the map $H^k(\mathbb{E}, \mathbb{Z}) \to H^{k+n}(E, E_o, \mathbb{Z}) : y \to y \cup u$ is an isomorphism( where $\cup$ is the cup product).

REMARK

Consider $\pi: E \to B$. Then the map $\varphi: H^k(B, \mathbb{Z}) \to H^{n+k}(E, E_o, \mathbb{Z})$ given by

$x \to (x^*, x) \cup u$ is an isomorphism, called the Thom isomorphism.

The above theorem enables us to introduce a characteristic class of $\mathbb{E}$. Consider the restriction homomorphism: $H^*(E, E_o, \mathbb{Z}) \to H^*(E, \mathbb{Z})$ induced by the inclusion $(E, Q) \hookrightarrow (E, E_o)$. The fundamental class $u \in H^n(E, E_o, \mathbb{Z})$ ( determining the orientation) determines a class $u \mid E \in H^n(E, \mathbb{Z})$. But $H^n(E, \mathbb{Z}) \simeq H^n(B, \mathbb{Z})$.

DEFINITION

The Euler class of an oriented n-bundle $\mathbb{E}$ is $e(\mathbb{E}) \in H^n(B, \mathbb{Z})$ which corresponds to $u \mid E \in H^n(B, \mathbb{Z})$ under $\pi^* : H^n(B, \mathbb{Z}) \to H^n(E, \mathbb{Z})$.

The following properties of Euler classes can be easily checked ( see eg. Milnor [602] ).

(i) If $f : B \to B'$ is covered by an orientation preserving bundle map, then $e(\mathbb{E}) = f^*(\mathbb{E}')$. If $\mathbb{E}$ is trivial, $n > 0$, then $e(\mathbb{E}) = 0$.

(ii) $e(\mathbb{E})$ changes its sign if the orientation is reversed.

(iii) If n is odd then $e(\mathbb{E}) + e(\mathbb{E}) \cdot = 0$.

(iv) The canonical map : $H^n(B, \mathbb{Z}) \to H^n(B, \mathbb{Z}_2)$ sends $e(\mathbb{E}) \to \omega_n(\mathbb{E})$.

(v) (a) $e(\mathbb{E} \oplus \mathbb{E}') = e(\mathbb{E}) \cup e(\mathbb{E}')$,

(b) $e(\mathbb{E} \times \mathbb{E}') = e(\mathbb{E}) \times e(\mathbb{E})$.

(vi) If $\mathbb{E}$ is oriented and admits a nowhere zero section then $e(\mathbb{E}) = 0$.

THEOREM 7

Let M be a compact oriented manifold. Then $\langle e(T_m), \mu \rangle = \chi(M)$ = the Euler characteristic (using rational or integeral coefficients). In the case of non-orientable manifolds $\langle \omega_n(\tau_m), \mu \rangle = \omega_n[M] = \chi(M) \bmod 2$.

## III. Chern numbers and Pontrjagin numbers

Let k be an integer. A partition for k is an un-ordered sequence $I = i_1, \ldots, i_r$ with $\Sigma i_t = k$. If $I = i_1, \ldots, i_r$, $J = j_1, \ldots, j_s$ are partitions for k and l respectively, then $IJ = i_1 \ldots i_r j_1 \ldots j_r$ is a partition for $k + 1$. If $I = i_1 \ldots i_r$ is a partition for k, then by a refinement we mean a product $I_1 \ldots I_t$ where $I_t$ is a partition for $i_t$.

### Chern Numbers

Let M be a compact complex manifold of dimension n. $I = i_1 \ldots i_r$ is a partition for n. The I th Chern number is defined by

$$C_I(M) = C_{i_1} \ldots C_{i_r}[M] = \langle C_{i_1}(\tau^n) \ldots C_{i_r}(\tau^n), \mu_{2n} \rangle ,$$

where $\tau^n$ is the tangent bundle of M, and $\Gamma_{2n}$ is the fundamental homology class determined by the canonical orientation of M.

Convention $C_I(M) = 0$ if I is a partition for some number other than n.

The Chern number of a compact Riemann surface is its Euler characteristic.

A manifild of dimension n has in general p(n) Chern numbers (p(n) = number of partition for n). These partitions are linearly independent in the sense there is no linear relation satisfied between them for all compact complex manifolds of dim n (for proof see Milnor [602], page 193).

Pontrjagin Numbers

Let M be a compact oriented manifold of dimension 4n. For $I = i_1 \ldots i_r$ a partition of n, the I th Pontrjagin number $P_I[M] = \langle p_{i_1}(\tau^{4n}) \ldots p_{i_r}(\tau^{4n}), \mu_{4n} \rangle$.

If we reverse the orientation of M, the Pontrjagin classes remain unchanged but $\mu_{4n}$ undergoes a sign change. Hence each Pontrjagin number $p_{i_1} \ldots {}_{i_r}[M]$ also changes sign. Thus if some Pontrjagin number $p_{i_1} \ldots {}_{i_r}[M]$ is non-zero then M does not admit an orientation reversing diffeomorphism.

Pontrjagin numbers are very useful in studying Thom's cobordism groups. In particular we have:

THEOREM 8

If a 4r-dimensional manifold M is the boundary of a compact differentiable oriented (4r + 1) - dimensional manifold, then every Pontrjagin number $p_{i_1} \ldots {}_{i_s}[M]$ is zero.

Another important property is that the signature (the index) of a compact differentiable oriented manifold of dimension 4r, can be expressed as a linear function of its Pontrjagin numbers (see Hirzebruch [379]).

For known results on the index of a 4-dimensional manifold see I-1-A-8 of this volume.

IV. Chern Classes

Chern Classes are topological cohomology classes in the base manifold and they measure how far a given bundle deviates from a product bundle. If $E \to X$ is a $C^\infty$ complex vector bundle over the $C^\infty$ manifold X, of fibre dimension m, the definition of its Chern classes $C_r(E)$, r = 1,2,m, are given such that $C_r(E) \in H^{2r}(X, R)$. We prefer to follow Chern-Weil and define these classes in terms of connections.

DEFINITION ( Horizontal Subspaces)

Let $P = P(X,G)$ be a differentiable principal bundle over a differentiable manifold with structure group G, a Lie group. Let $p : P \to X$ be the projection. For $u \in P$, let $F_u$ be the fibre through u. Let TP be the tangent bundle of P and $T_u(F_u)$ be the space of tangent vectors to $P_u$ at u. Suppose in $T_u(P)$ we can choose a space $T_{\underline{u}}(P)$ such that:

(a) $T_u(P)$ is the direct sum of $T_u(P)$ and $T_u(F)$.

(b) The correspondence $u \to T_u(P)$ is invariant under G.

(c) $T_{\underline{u}}(P)$ depends differentiably on u.

$T_{\underline{u}}(P)$ is called a horizontal subspace of $T_u(P)$ at u. $T_u(F_u)$ is called the vertical subspace of $T_u(P)$ at u. Given $T_u(F_u)$ and $T_{\underline{u}}(P)$ every vector $X_u \in T_u(P)$ admits a unique decomposition $X_u = {}^{\perp}X_u + X_{\underline{u}}$ where ${}^{\perp}X_u \in T_u(F_u)$ and $X_{\underline{u}} = T_{\underline{u}}(P)$.

DEFINITION

Given a principal bundle $P = P(X,G)$, if we can define a field of horizontal subspaces in $T(P)$ we say we have a connection $\Gamma$ in P.

Given a connection $\Gamma$ in P, we associate with it a 1-form $\omega$ on P with values in the Lie algebra $\underline{G}$ of G as follows.

For each vector field $X \in T(P)$, we define $\omega(X)$ to be the unique $A \in \underline{G}$ such that $A^* = {}^{\perp}X$. (G acts on P on the right; consider $A \in \underline{G}$. Associate to A the set $a_t = \exp tA$, t real: this is a 1-parameter subgroup of G and its action on P induces a vector field $A^*$.) The form $\omega$ is called the connection form of $\Gamma$. Let $\Theta$ be the canonical 1-form on G, ie., $\Theta$ is the left invariant $\underline{G}$- valued 1-form satisfying $\Theta(A) = A$, for every $A \in \underline{G}$.

DEFINITION (Covariant Differentiation)

Given a connection $\Gamma$ in $P = P(X,G)$, we can associate to the exterior differentiation operator d an operator $\nabla$ as follows. For any $\alpha$ on P, $\nabla\alpha = (d\alpha)h$ where h is the projection $T_u(P) \to T_{\underline{u}}(P)$. $\nabla$ is called the exterior covariant differentiation and the form $\nabla\alpha$ is called the exterior covariant derivative of $\alpha$.

DEFINITION ( Curvature and Torsion of a Connection)

Let a connection $\Gamma$ be given in $P = P(X,G)$: let the connection form $\omega$ and the canonical form on G be $\Theta$. Set $\Omega = \nabla\omega$, $\Theta = \nabla\Theta$. $\Omega$ is called the curvature form of the connection and $\Theta$ is called the torsion form of the connection.

Relating these we have the structure equation.

THEOREM 9

$$d\omega = -1/2\ [\omega,\omega] + \Omega .$$

Chern-Weil homomorphisms

Let G be a Lie group with Lie algebra $\underline{G}$. Consider the space $I^k(G)$ of k-linear symmetric maps

$$\underline{G} \times \ldots \times \underline{G} \to \mathbb{R} \quad (\text{k factors})$$

which are invariant under G, ie., for every $a \in G$ and for every

$(t_1 \ldots y_k) \in \underline{G} \times \ldots \times \underline{G}$, $f((\mathrm{ada})T_1 \ldots (\mathrm{ada})t_k) = f(T_1 \ldots t_k)$.

Set $I(G) = \sum_{k \geq 0} I^k(G)$.

$I(G)$ can be made into a commutative algebra by defining multiplication in a familiar way.

Let now $P = P(X,G)$ be a principal bundle with a given connection $\Gamma$. Define for $f \in I^k(G)$ and $X_i \in T_u(P)$, $1 \leq i \leq 2k$:

$$F(\Omega)(X_1, \ldots, X_{2k}) = \frac{1}{(2k)!} \Sigma \pm f(\Omega(X_{\sigma(1)}, X_{\sigma(2)}) \ldots \Omega(X_{\sigma(2k-1)}, X_{\sigma(2k)}))$$

where the summation is taken over all permutations of $\sigma$ of $(1,2,\ldots,2k)$. $F(\Omega)$ is a 2k-form on P.

With these preliminary definitions and notations we state a basic theorem in the theory of characteristic classes.

THEOREM 10 (Chern-Weil)

Let $P = P(X,G)$ be a principal bundle on X, with group G and projection map P. Let $\Gamma$ be a connection in P and $\Omega$ its curvature form. Then:

(a) For each $f \in X^k(G)$, the 2k-form $F(\Omega)$ on P projects into a unique closed 2k form, say $\gamma(\Omega)$ on X, ie, $F(\Omega) = p^*(\gamma(\Omega))$.

(b) Let $W(f)$ be the element of the de Rham cohomology group $H^{2k}(X,R)$ defined by the closed 2k-form $\gamma(\Omega)$. Then $W(f)$ is independent of the choice of the connection $\Gamma$ and $W : I(G) \to H^*(X,R)$ is an algebra homomorphism.

REMARK

W is called the Chern -Weil homomorphism. The image $W(S(G))$ is the algebra of characteristic classes of P. For the proof of the theorem see e.g. Kobayashi-Nomizu [475], Volume 2. For a proof of the theorem using universal connections see Narasimhan-Ramanan [677].

DEFINITION ( Chern Classes)

Let E be a complex vector bundle over X with fibre $C^m$ and group Gl(m, C) and let P be its associated principal bundle. Given a connection $\Gamma$ on P with its curvature form $\Omega$, define polynomial functions $f_1 \ldots f_m$ on the Lie algebra $\underline{G}l$ (m, C) by

$$\det(\lambda I_m - (\frac{1}{2\sqrt{-1}\pi})\Omega) = \sum_{0 \leq k \leq m} f_k(\Omega)\lambda^{m-k} .$$

These are invariant under ad(Gl(m, C)). Hence by the above theorem of Chern-Weil, there exists for each k, $0 \leq k \leq m$ a unique closed 2-form $\gamma_k$ on X such that $p^*(\gamma_k) = f_k(\Omega)$, where $p: P \to X$ is the projection, so that we can write

$$\det(I_m - (\frac{1}{2\sqrt{-1}\pi})\Omega) = p^*(1 + \gamma_1 + \ldots + \gamma_m) .$$

DEFINITION

The k th Chern class $C_k(E)$ of E is represented by the 2k-closed forms $\gamma_k$ ( and thus $C_k(E) \in H^{2k}(X,R)$).

$C(E) = C_o(E) + \ldots + C_m(E)$ is called the total Chern class of E.

The following theorem shows that Chern classes are the obstructions for a bundle to be trivial.

THEOREM 11

Let $E \to X$ be a differentiable vector bundle of rank m. Then

(a) $C_o(E) = 1$.

(b) If $E \cong X \times C^m$ ( trivial), $C_j(E) = 0$, for $j = 1 \ldots m$.

(c) If $E \cong F + F'$ where F' is trivial bundle of rank r, $C_j(E) = 0$ for $j = m-r+1, \ldots, M$.

The following is the theorem of Bott referred to in Part B Section 1 of Chapter 1.

THEOREM 12 (Bott)

Let E be any complex vector bundle of rank n over the sphere $S^{2n}$. Then $C_n(E).S^{2n}$ is divisible by $(n-1)!$ ( factorials).

REMARKS

(1) $C_n(E).S^{2n}$ stands for the value of $C_n(E)$ over the fundamental cycle $S^{2n}$.

(2) Bott's theorem is an integrability theorem. For a more general integrability theorem see Schwarzenberger [823], [379] - Appendix. (3) In analogy with Stiefel-Whitney classes, Chern classes can also be defined axiomatically. Conditions (i) to (v) are satisfied. In (v) the canonical line bundle over $P^1(C)$ must be taken. The Chern classes are uniquely determined by these conditions. (4) It can be shown that the Chern classes are topological invariants: they depend only on the underlying topological bundle.

## V. Pontrjagin Classes

Let $E \to X$ be a differentiable $Gl(n,R)$ bundle. For r a positive integer, let $p_{r/2}$ be the homogeneous polynomial of degree r which is the coefficient of $\lambda^{n-r}$ in $\det(\lambda - 1/2\pi.A) = \Sigma p_{r/2}(A,,,,A)\lambda^{n-r}$, $A \in \underline{G}l(n,R)$. Then $p_{r/2} \in I^r(Gl(n,R))$: The Chern-Weil images of $W(p_{r/2})$ are called the Pontrjagin classes.

Thus $p_{r/2}(E) \in H^{2r}(X,R)$, $r = 0,\ldots,n$. Note $p_{r/2}(E) = 0$ for r odd.
$p(E) = p_o(E) + \ldots + p_{[n/2]}(E) \in H^*(X,R)$ is called the total Pontrjagin class.

THEOREM 13

Let $\mathbb{E} : E \to X$ be a differentiable $Gl(n,R)$ bundle. Then

(i) $p_r(E) \in H^{4r}(X,R)$, $r = 0,1,\ldots$ .
$p_o(E) = 1$ and $p_r(E) = 0$ for $r > n/2$.

(ii) Let $\mathbb{E}_C : E_C \to X$ be the complexification of $\mathbb{E}$ (which is a $Gl(n.C)$ bundle). Then $p_r(E) = (-1)^r C_{2r}(E_C)$ $r = 0,1,\ldots$ .

(iii) If $f : Y \to X$ is a continuous map and $\mathbb{E} : E \to X$ is a $Gl(n,R)$ bundle, then $p f^*(E) = f^* p(E)$. (Naturality).

(iv) If $\mathbb{E} : E \to X$ is a $Gl(n,R)$ bundle and $\eta : F \to X$ is a $Gl(m,R)$ bundle then $p(E \oplus F) = p(E).p(F)$. (Whitney Product Theorem). That is $p_k(E \oplus F) = \sum_{i+j=r} p_1(E) \cup P_j(F)$, $r = 0,1,\ldots,[\frac{n+m}{2}]$.

Thom introduced Pontrjagin classes $p_i \in H^{4i}(M,Q)$ for PL manifolds and showed that these are PL invariants ( see Milnor [602], Section 20). In 1965, Novikov [704] proved a deep theorem showing that these are in fact topological invariants.

For some recent developments in characteristic classes see Chern-Simons [158], Chern [156].

## SECTION 2

## The Problem of Uniqueness of Complex Structures on the Complex Projective Spaces

### I-2-1- (Complex Projective Spaces)

Let us first recall the definition of complex projective spaces. Define an equivalence relation in $C^{n+1} - \{0\}$ by declaring two points to be equivalent if their coordinates differ from each other by a non-zero complex number and denote the resulting quotient space by $P^n$. There is the natural projection map $\pi : C^{n+1} - \{0\} \to P^n$ and $P^n$ is a Hausdorff space in the quotient topology. $P^n$ is compact since $\pi \mid S^n : S^n \to P^n$ is a continuous surjective map. If $Z = (Z^o, \ldots, Z^n) \in C^{n+1} - \{0\}$, denote $\pi(Z)$ by $[Z]$ and call $(Z^o, \ldots, Z^n)$ the homogeneous coordinates of $[Z]$.

If $(w^o, \ldots, w^n)$ is another set of homogeneous coordinates for $[Z]$, from the definition it is clear that there exists a non-zero complex number $\lambda$ such that $w^r = \lambda Z^r$, $0 \leq r \leq n$. We define a complex structure on $P^n$ as follows. Let $U_j = \{p \in P^n \mid \text{the } j^{th} \text{ homogeneous coordinate of p is not zero}\}$. $U_j$ forms an open covering of $P^n$.

Define $h_j : U \to C^n$ by

$$h_j([Z]) = \left( \frac{Z^o}{Z^j}, \ldots, \frac{Z^{j-1}}{Z^j}, \frac{Z^{j+1}}{Z^j}, \ldots, \frac{Z^n}{Z^j} \right).$$

Then $h_j$ is easily seen to be a homeomorphism onto $C^n$ and $h_j \circ h_k^{-1}$ is a biholomorphic map, for each j,k such that

$$U_j \cap U_k \neq 0.$$

$P^n$ with this complex structure is denoted by $P^n(C)$. $P^n(C)$ is a compact connected complex n-manifold.

The fundamental question is whether this is the only way of prescribing a complex structure on $P^n$. Until now no other complex structure is found to exist on $P^n$. The longstanding conjecture is:

CONJECTURE : The complex structure on $P^n$ is unique.

It is easily checked that the complex structure on $P^1$ is unique.

A number of important partial results are known relating to this conjecture. We state these important results in this section.

I-2-2- (The theorem of Hirzebruch-Kodaira)

A hermitian metric on a complex manifold M with local coordinates $(Z^n_j)$ is given by

$$d\,S^2 = \sum_{\alpha,\beta=1}^{n} g_{j\alpha\overline{\beta}}\,(Z)\; d\,Z^{\alpha}_j \;\; d\,\overline{Z}^{\beta}_j$$

where $g_{j\alpha\overline{\beta}}\,(Z)$ is a $C^\infty$ section of $T^* \otimes \overline{T}^*$ such that

(i) $\overline{g}_{j\;\alpha\overline{\beta}}\,(Z) = g_{j\beta\overline{\alpha}}\,(Z)$ and

(ii) $\sum_{\alpha,\beta=1}^{n} g_{j\alpha\overline{\beta}}\,(Z)\,\xi^{\alpha}\,\overline{\xi}^{\beta} \geq 0$ with equality if and only if $\xi = 0$, where

$T^*$ denotes the dual of the tangent bundle T of M.

Using partitions of unity we can introduce a hermitian metric on any complex manifold. We associate to the hermitian metric a differential form $\omega$ of type (1,1) given by $\omega = \sqrt{-1}\,\Sigma\, g_{j\alpha\beta}\, dZ^{\alpha}_j\, dZ^{\beta}_j$. This form $\omega$ is a real form. In general it is not d-closed where d denotes the exterior derivative. If a hermitian metric is such that the associated form $\omega$ is d-closed, it is called a Kahler metric and $\omega$ a Kahler form. A complex manifold is Kahler if there is a Kahler metric on it. It is easy to check that any submanifold of a Kahler manifold is Kahler. Every compact Riemann surface R (compact connected complex 1-manifold) is Kahler since any metric on R is Kahler. However not every compact complex manifold is Kahler. There are topological restrictions for a complex manifold to admit a Kahler metric. Necessarily for a Kahler manifold the odd Betti numbers must be even. A conjecture of Kodaira is that for a compact complex surface to be Kahler it is sufficient that the first Betti number is even (page 85, Kodaira-Morrow [489]). This conjecture is true for a large class of compact complex surfaces (Miyaoka [606]).

$P^n(C)$ are Kahler (the Fubini-Study metric on $P^n(C)$ is Kahler). A deep theorem of Hirzebruch-Kodaira [386] states that the only Kahler complex structure on $P^n(C)$ when n is odd is this standard one. When n is even, only if further conditions are satisfied, do we have a similar assertion. Their theorem is

THEOREM 1 (Hirzebruch-Kodaira)

Let M be a compact connected Kahler complex manifold of dimension n such that M is diffeomorphic to $P^n(C)$. Then

(a) If n is odd, M is biholomorphic with $P^n(C)$

(b) Let n be even; let $g \in H^2(M,Z)$ generate $H^2(M,Z)$ and belong to the cohomology class of Kahler metric on M. If $C_1(M) \neq -(n+1)g$, then M is biholomorphic with $P^n(C)$. ($C_1(M)$ denotes the first Chern class of M.)

Remark

There is a similar theorem for n-dimensional quadrics in $P^{n+1}(C)$, (see Brieskorn [116]).

We can ask whether in the above theorem the assumption M is diffeomorphic to $P^n(C)$ could be replaced by M is homeomorphic to $P^n(C)$. Using the theorem of Novikov [704] that the rational Pontrjagin classes are topological invariants this can be done. For details see Morrow [625]. Very recently S. T. Yau has shown that if a compact complex surface is homeomorphic to $P^2(C)$, then it is biholomorphic to $P^2(C)$. He has proved this deep result as an application of his solution of the famous Calabi Conjecture: Yau's solution of Calabi's conjecture is acclaimed to be a great achievement of recent times.

I-2-3- (Calabi's Conjecture)

Let M be a Kahler manifold with Kahler metric $dS^2$. The Ricci tensor of the metric with respect to local coordinates can be expressed as $\sum_{\alpha,\beta} R_{j\alpha\beta}\, dZ_j^{\alpha}\, dZ_j^{\beta}$. After computation we get $R_{j\alpha\overline{\beta}} = -\frac{\partial^2}{\partial Z^{\alpha}\partial \overline{Z}^{\beta}} \log\det(g_{j\alpha\overline{\beta}})$.

The (1,1) form $(\sqrt{-1}/2\pi)\sum_{\alpha,\beta} R_{j\alpha\overline{\beta}}\, dZ^{\alpha} \wedge d\overline{Z}^{\beta}$ is called the Ricci form of the Kahler metric $dS^2$. It is easy to check that it is closed. Chern [151] has shown that its cohomology class is the first Chern class $C_1(M)$ of M. Thus a necessary condition for a closed (1,1) form $\omega$ on M to be the Ricci form of some Kahler metric on M is that its cohomology class $[\omega]$ must represent the first Chern class $C_1(M)$.

Calabi's conjecture [134] is that this condition is also sufficient. In 1977, Yau [987] solved this conjecture in the affirmative. For details see Yau [989].

Yau reduced the conjecture to a problem in nonlinear partial differential equations and then using his earlier work on the Monge-Ampere equation settled the Calabi conjecture in the affirmative. There are many applications of this very deep work; one application being that the complex structure of the complex projective plane is unique.

Subsequent to Yau's solutions of the Calabi conjecture, Aubin [66] and Bourguignon [113] have given different proofs of the solution. See also Kazdan [452].

1-2-4- (Frankel Conjecture)

The problem of the uniqueness of a complex structure on $P^n(C)$ can also be studied by studying its curvature properties. Let us recall first some well known definitions.

Let $(M,g)$ be a compact Kahler n-manifold. M can be regarded as a Riemannian manifold $\underline{M}$. Let R be its Riemannian curvature tensor. The sectional curvature K of $\underline{M}$ is defined on the Grasmann bundle of 2 planes of the tangent bundle of $\underline{M}$. Choose a plane $\sigma$ in the tangent space $T_x(\underline{M})$. Then with respect to an orthonormal basis X,Y for $\sigma$, the sectional curvature $K(\sigma) = R\ (X,Y,X,Y)$. In general $K(\sigma)$ depends on x. If $K(\sigma)$ is constant for all x and for all $\sigma$, then $\underline{M}$ is of constant sectional curvature. Assume that $\sigma$ is invariant under the (almost) complex structure J of M. The set of J invariant planes is a holomorphic bundle over M with fibre $P^{n-1}(C)$. The restriction of the sectional curvature function K to this bundle is called the holomorphic sectional curvature $H(\sigma)$. Thus $H(\sigma)$ is defined if $\sigma$ is J-invariant and $H(\sigma) = K(\sigma)$. If $H(\sigma)$ is constant for all x and for all J-invariant $\sigma$, we say M is of constant holomorphic sectional curvature.

Given two J-invariant planes $\sigma_1, \sigma_2$ define

$$H(\sigma_1, \sigma_2) = R(X, JX, Y, JY),$$

where X is a unit vector in $\sigma_1$ and Y is a unit vector in $\sigma_2$. It can be checked that $R(X, JX, Y, JY)$ depends only on $\sigma_1, \sigma_2$. $H(\sigma_1, \sigma_2)$ is called the holomorphic bisectional curvature. This notion was introduced by Goldberg and Kobayashi [282] in 1967. If $H(\sigma_1, \sigma_2)$ is constant for all points x and for all J-invariant planes $\sigma_1, \sigma_2$,

M is said to be of constant holomorphic bisectional curvature. Since $H(\sigma,\sigma) = H(\sigma)$, holomorphic bisectional curvature carries more information than the sectional curvature does, and hence the importance of this new notion.

By applying the Bianchi identity, we see that $H(\sigma_1,\sigma_2) = R(X,Y,X,Y) + R(X,JY,X,JY)$ and therefore $H(\sigma_1,\sigma_2)$ is a sum of two holomorphic sectional curvatures essentially. In particular, if a manifold is of positive holomorphic sectional curvature, then it is of positive holomorphic bisectional curvature. In complex dimension 2 everywhere positive (non-negative) bisectional curvature is equivalent to everywhere positive (non-negative) curvature.

It is easy to check that the $P^n(C)$ with the Fubini-Study metric is of constant sectional curvature (which can be normalized to be 1) and hence $P^n(C)$ has constant holomorphic bisectional curvature. In 1961, Frankel [250] proved that if M is a compact connected Kahler surface of positive sectional curvature, then it is biholomorphic to $P^2(C)$. He conjectured that this remains true in higher dimensions also.

Frankel Conjecture: A compact connected complex n-manifold which admits a Kahler metric of positive sectional curvature is biholomorphic to $P^n(C)$.

For $n = 1$, this is trivially true. Kobayashi-Ochiai proved the Frankel conjecture for $n = 3$ (see, however, the remark of Ochiai on page 120 in Ochiai [707]). However very recently Mori [620] has proved a more general conjecture of Hartshorne which settles the Frankel conjecture in the affirmative for all n. Before stating this conjecture of Hartshorne we mention a slightly more general conjecture than Frankel's.

Goldberg-Kobayashi-Ochiai Conjecture: A compact connectec complex n-manifold which admits a Kahler metric of positive bisectional curvature is biholomorphic to $P^n(C)$.

For $n = 1,2$ this conjecture is true (as we remarked earlier for $n = 2$, everywhere positivity of bisectional curvature implies everywhere positivity of sectional curvature). Under an additional assumption, Kobayashi-Ochiai proved the conjecture for $n = 3$. It is clear that validity of Frankel's conjecture implies that of the Goldberg-Kobayashi-Ochiai conjecture.

If one imposes further conditions on the curvature, then better results are available. For example, Berger proved that any compact connected Kahler-Einstein n-manifold with everywhere positive sectional curvature is holomorphically isometric with $P^n(C)$ (with its Fubini-Study metric). This is also true in the case of everywhere positive bisectional curvature (Goldberg-Kobayashi [282]). The assumption also holds for compact connected Kahler n-manifolds of constant scalar curvature and everywhere positive bisectional curvature.

I-2-5- (The pinching characterization of $P^n(C)$)

Let $(M,g)$ be a compact connected Kahler n-manifold. Let $\sigma$ be any plane in the tangent space $T_x(\underline{M})$ at $x \in M$. The Kahler sectional curvature of $\sigma$ is defined by

$$H^*(\sigma) = \frac{4K(\sigma)}{1 + 3g(X,JY)^2}$$

where $J$ is the complex structure tensor, $X,Y$ form an orthonormal basis for $T_x(\underline{M})$. If $\sigma$ is J-invariant $H^*(\sigma)$ reduces to $H(\sigma)$.

Let $\delta$ be a real number such that $0 < \delta \leq 1$. Suppose there exists a constant $A$ such that, for all planes $\sigma$ in $T_x(\underline{M})$ and all $x \in M$, $\delta A \leq K(\sigma) \leq A$. Then we say that the Kahler manifold $(M,g)$ is $\delta$-pinched (or positively $\delta$-pinched). We say $(M,g)$ is $\delta$-Kahler pinched (or positively $\delta$-Kahler pinched) if there exists a constant $A$ such that $\delta A \leq H^*(\sigma) \leq A$, for all planes $\sigma$ in $T_x(\underline{M})$ and for all $x \in M$. Similarly, $(M,g)$ is called $\delta$-holomorphically pinched (or positively $\delta$-holomorphically pinched) if there exists a constant $A$ such that

$$A \leq H^*(\sigma) \leq A,$$

for all planes $\sigma$ in $T_x(M)$ invariant under $J$ and for all $x \in M$. In general, a complex manifold need not carry a pinched Kahler metric. There are a number of implications between these pinchings (if they exist) which we shall not go into. We can get information on the homotopy type of Kahler manifolds from the pinching information. For example, a compact connected Kahler n-manifold which is $\delta$-pinched with $\delta > (9/16)$ is of the same homotopy type as $P^n(C)$ (Klingenberg [467]). We have the following important theorems. See Kobayashi-Nomizu [475].

THEOREM 2

Suppose a compact connected Kahler n-manifold $(M,g)$ satisfies any one of the three conditions (i) $M$ is $(1/4)$-pinched (ii) $M$ is 1-holomorphically pinched (iii) $M$ is 1-Kahler pinched. Then $M$ is holomorphically isometric with $P^n(C)$ (with

Fubini - Study metric ).

For further results on Pinched manifolds see Berger [78], Bishop-Goldberg [88], Klingenberg [467], Kabayashi [471], Tsagas [927].

I-2-6- (Hartshorne's Conjecture)

Ample Line Bundles: Let M be a compact connected complex n-manifold. A holomorphic line bundle $L \to M$ is said to be generated by its sections if given any point $x \in M$, there exists a holomorphic section $s \in H^o(M, \underline{L})$ such that $s(x) \neq 0$. This is equivalent to requiring that the evaluation map $H^o(M, \underline{L}) \to L_x$, where $L_x$ is the fibre at x, is surjective for all $x \in M$. Let $P(H^o(M, \underline{L})^*)$ denote the projective space associated to the dual of $H^o(M, \underline{L})$.

If $L \to M$ is generated by its sections, there exists a canonical map $\Phi_L : M \to P(H^o(M, \underline{L})^*)$ defined as follows: let $N + 1 = \dim_C H^o(M, \underline{L})$. For each $x \in M$, let $H^o(M, \underline{L})_x$ denote the subspace of sections vanishing at x. Then $\dim H^o(M, \underline{L})_x = N$. Define $\Phi_L(x)$ to be the subspace of $H^o(M, \underline{L})$ orthogonal to $H^o(M, \underline{L})_x$. Since $L_x^* \to H(M, \underline{L})^*$ is injective, this map is well defined. Choose a basis $s_o, \ldots, s_N$ for $H^o(M, \underline{L})$. Then locally $\Phi_L$ is given by $x \to (f_o(Z), \ldots, f_N(Z))$, $f_i$ are holomorphic, (in terms of homogeneous coordinates with respect to the dual basis of $H^o(M, \underline{L})$). Hence $\Phi_L$ is holomorphic. If, further, $\Phi_L$ is an embedding ($\Phi_L$ and $d\Phi_L$ are both injective), we say that the given line bundle L is very ample. L is said to be ample if for some integer k, $L \otimes \ldots \otimes L$ (k times) is very ample.

Ample Vector Bundles: Let $E \to M$ be a holomorphic vector bundle of rank r. Associated to this there exists a holomorphic fibre bundle $IP(E) \to M$, called the projective bundle of E which is defined as follows. Let $s_o$ denote the zero section of E (it maps each $x \in M$ into the zero of the fibre $E_x$ at x). The group $C^*$ of nonzero complex numbers acts on $E - s_o(M)$ and the quotient $P(E) = E - s_o(M)/C^*$ has the structure of holomorphic fibre bundle over M with group $C^*$ and the fibre at any point $x \in M$ is the projective space $P(E_x)$ of lines in the vector space $E_x$. The following two properties can easily be checked.

(i) If two holomorphic bundles $E_1, E_2$ are such that their associated projective bundles are bundle isomorphic, then there exists a line bundle L such that $E_1 = E_2 \otimes L$, and conversely.

(ii) A holomorphic line subbundle of E gives rise to a holomorphic section of IP(E) and, conversely, every section of IP(E) arises this way.

Let $E \to M$ be a holomorphic vector bundle and $IP(E) \to M$ be the associated projective bundles. Consider the pull back bundle $\pi^* E \to IP(E)$. Consider the holomorphic line subbundle of this bundle, whose fibre at any point $(v,x) \in IP(E)$ is the line in the vector space $E_x$ represented by v. This particular holomorphic line bundle plays an important role; it is called the tautological line bundle on IP(E) and is denoted by $\Theta_{IP(E)}(1) \to IP(E)$.

We say that a holomorphic vector bundle $E \to M$ is ample (very ample) if the tautological line bundle $\Theta_{IP(E)}(1)$ is ample (very ample). The following important theorem of Hartshorne gives equivalent conditions:

THEOREM 3

Let $E \to M$ be a holomorphic vector bundle over a compact connected complex manifold. Then the following statements are equivalent.

(a) E is ample. (b) For every coherent sheaf F on M, the cohomology groups $H^i(M, F \otimes S^n(E)) = 0$ for all i and sufficiently large n where $S^n(E)$ denotes the n-fold symmetric product of E.

(For more on ample vector bundles see Hartshorne [354], [356] and Griffiths [318]).

The tangent bundles of $P^1(C)$ and $P^2(C)$ are ample and Hartshorne [354] has shown these are the only compact complex manifolds in dimension 1 and 2, respectively, whose x tangent bundles are ample. He conjectured that this must be true for any n. All the definitions and discussions of this section, though stated for the complex field C, remain valid in any algebraically closed field of any characteristic.

Hartshorne's Conjecture: Every irreducible n-dimensional projective variety with ample tangent bundle defined over an algebraically closed field k (of characteristic $\geq 0$) is isomorphic with the n dimensional complex projective space $P^n(k)$ over k.

Very recently Mori [620] has settled this conjecture in the affirmative. When the field k is C, it is known that ampleness of the tangent bundle is implied by the positivity of the holomorphic sectional curvature (e.g. Griffiths [319]). Hence Mori's proof of Hartshorne's conjecture implies that Frankel's conjecture is true (and hence the Goldberg-Kobayashi-Ochiai conjecture is also true).

Regarding the importance of the Calabi conjecture and the Hartshorne conjecture, it suffices to state that the problems are mentioned in the 'Problems of Present Day Mathematics' in Browder [126]. The problems 1, 1a and 6 mentioned on pages 51,53, in Browder [126] are now settled in the affirmative.

I-2-7- (Characterization of $P^n(C)$ by its Spectrum)

> "The study of eigenvalues of differential operators is one of the most important and far reaching problems in mathematics. It is fundamental in mechanics, physics, geometry and it has of course been extensively worked on from many different points of view."
>
> M.F. ATIYAH.

In recent yearsm especially since Kac's provacative paper "Can one hear the shape of a drum?" Amer. Maths. Monthly 73, 1 - 23 (1966), significant developments have taken place in relating the geometry of Riemannian manifold to the spectrum of its Laplacian. Here again we have witnessed the confluence of many branches of mathematics giving vital information on fundamental problems of mathematics and physics.

Let $(M,g)$ be a compact connected Riemannian manifold with Riemannian metric g. One way of defining the Laplacian $\Delta$ acting on differentiable $(c^\infty)$ functions $f: M \to R$ is as follows: let d be the exterior derivation and D be the covariant differentiation operator. Then $-\Delta f$ is defined to be the trace of the 2-form $Ddf$ with respect to the given metric g. In terms of local coordinates $(x_i)$,

$$\Delta f = \left(\frac{-1}{g}\right) \sum_{i,j} \frac{\partial\left(g g^{ij}\left(\frac{\partial f}{\partial x j}\right)\right)}{\partial x_i} \quad \text{with } g = \det(g_{ij}).$$

In $R^n$ this reduces to the usual Laplacian with its sign changed. The Laplacian $\Delta$ on the manifold is intrinsically defined: it depends only on the metric g. $\Delta$ is a positive self-adjoint elliptic operator. Since $\Delta$ is self-adjoint there exists a spectral resolution of $\Delta$ into a complete orthonormal set of eigenfunctions and the corresponding eigenvalues. The spectrum of $(M,g)$ (also called the spectrum of $\Delta$) denoted by Spec $(M,g)$ is the set $\{\lambda \in \mathbb{R} \mid \quad f: M \to R,\ f \neq 0$, such that $\Delta f = \lambda f\}$. Since $\Delta$ is elliptic the spectrum is discrete and the multiplicity of each eigenvalue must be finite. Zero is an eigenvalue and its multiplicity is one since $\Delta f = 0 = f$ is constant.

There can be no negative eigenvalue since $\Delta$ is a positive operator. Hence

$$\text{Spec}(M,g) = \{0 = \lambda_o < \lambda_1 \le \lambda_2 \le \dots\},$$

where $\lambda_1$, $i > 1$ is repeated as many times as its multiplicity. If $\varphi$ are the corresponding eigenfunctions, then the series $\Sigma \exp(t\lambda_i)\varphi_i(x) \otimes \varphi_j(y)$ converge uniformly on compact subsets of $(0,\infty) \times M \times M$ to the fundamental solution of the Heat Equation $(\frac{\partial}{\partial t} + \Delta)\varphi = 0$.

It may be surprising that Spec $(M,g)$ has been computed, until now, only in very special cases. Spec $(S^n, g)$ is completely known when $S^n$ is the standard n-sphere with its standard Riemannian metric. In the case of $S^1$, $\Delta f = d^2 f/d\theta^2$, where $\theta$ represents arc length on $S^1$. The eigenfunctions are the usual trigonometric functions and the eigenvalues are $(n^2)$, $n \in \mathbb{Z}$, with multiplicity 2. That is Spec $(S^1,g) = \{0,1,1,4,4,9,9,\dots\}$. It is well known that any smooth function $S^1$ has a Fourier series expansion. For any n it is known that Spec $(S^n,g) = \{0 = < \lambda_o \le \lambda_1 \le \lambda_2 \dots\}$ where $\lambda_j = j(n+j-1)$ with multiplicity $\binom{n+j}{n} - \binom{n+j-2}{n}$. The eigenfunctions are given by spherical harmonics. See Hormander [411]. Spectrum of tori are also known. In general it is very difficult to complete the spectrum of a Riemannian manifold.

One can immediately pose a challenging question. Given any compact connected differentiable n-manifold, describe the image of the map:

$$\text{Spec}: \{\text{Riemannian Structures on } M\} \to \mathbb{R}^N.$$

"The problem of describing the image looks too formidable today even to talk about it" remarked Berger one day in 1975, see page 133, [79]. Inspite of further advancements in spectral geometry, Berger's assessment still continues to be valid. The above map Spec cannot be injective, as can be verified in the case of tori (see e.g. Berger [79]).

One interesting and important line of investigation is to determine to what extent the geometry of $(M,g)$ is reflected by its Spec $(M,g)$. Milnor [597] was the first to show that the spectrum may not completely determine the geometry. He produced two nonisometric compact connected 17-dimensional Riemannian manifolds with the same spectrum. Recently Donnelly [206] produced two compact connected Kahler manifolds having the same spectrum but they are not biholomorphic. Inspite of these examples, recent investigations have shown that significant geometrical properties of a Riemannian manifold are reflected by its spectrum. Patodi has shown that whether or not a Riemannian manifold has constant scalar curvature, constant sectional curvature or flat curvature can be determined by studying its spectrum. Gilkey has shown that by studying the spectrum of a Hermitian manifold it is possible to determine whether or not the manifold admits a Kahler metric, and it is possible to determine whether a given Kahler metric on the manifold has constant holomorphic sectional curvature. More importantly Patodi, quickly followed by Atiyah-Bott-Patodi and Gilkey, gave a purely analytic proof of the famous Atiyah-Singer index theorems.

The main tool in the study of the problems of the above type is the heat equation. Associated with any differential operator on a manifold there exists a heat equation; if the operator is nice, the fundamental solution to this heat equation exists. Then the asymptotic expansion of the fundamental solution links the geometry of the manifold with the spectrum of the manifold.

REMARK 1

We have given the definition of the Laplacian $\Delta$ acting on smooth functions. More generally the Laplacian $\Delta^p$ acting on smooth p-forms on M is defined by $\Delta^p = d\delta + \delta d$, where $\delta$ is the adjoint of d with respect to the chosen metric. (See e.g. Wells [961] for the definition of $\delta$). As earlier, there is $\text{Spec}\ p\ (M,g) = \{0 \leq \lambda_{o,p} < \lambda_{1,p} \leq \lambda_{2,p} \leq \ \ldots\}$. An outstanding problem here is whether the eigenvalues of the smooth functions on M determine the eigenvalues of the smooth p-forms for every p.

PROBLEM Does $\{\lambda_{n,o}\}$ determine $\{\lambda_{n,p}\}$ for every p? This is known to be true in dimension 2 and for rank one symmetric spaces.

REMARK 2

Still more generally one can define the Laplacian acting on smooth maps between Riemannian manifolds. The basic reference here is Eells-Sampson [225]. For an excellent account of recent developments in the theory of harmonic maps, see Eells-Lemaire [224].

REMARK 3

Smooth functions f (on forms) which are solutions of the corresponding Laplace equation $\square f = 0$ are called harmonic. It is a classical fact of the Hodge theory that the space of harmonic functions (or forms) is finite dimensional.

Now consider a compact connected complex n-manifold M with a hermitian metric g. Let $A^{p,q}(M) \to A^{p,q+1}(M)$ be the usual $\bar{\partial}$ operator and $\delta$ be its adjoint operator. Then the complex Laplacian $\square$ is defined by $\square = \bar{\partial}\delta + \delta\bar{\partial}$. $\square$ maps (p,q) forms into (p,q) forms. It is also an elliptic self-adjoint operator. As before, the spectrum of (M,g) is defined to be the set of eigenvalues

$$\text{Spec}(M,g) = \{0 \leq \lambda_o^{p,q} \leq \lambda_1^{p,q} \leq \dots\},$$

each eigenvalue being repeated as many times as its multiplicity. If $\varphi_i^{p,q}$ are the corresponding eigenfunctions, then $\Sigma \exp(t\lambda_i^{p,q})\,\varphi_i^{p,q}(Z) \otimes \varphi_i^{p,q}(Z')$ converge on compact subsets of $(0,\infty) \times M \times M$ to the fundamental solution of the heat equation $(\frac{\partial}{\partial t} + \square)\varphi = 0$.

Now M can be regarded as a Riemannian manifold, the Riemannian metric being the real part of the hermitian metric, and hence there is the real Laplacian $\Delta$. In general there may not be any relation between the complex Laplacian and the real Laplacian. But if M is a Kahler manifold it is wellknown that $\square = 2\Delta$. As already remarked, whether or not a Hermitian manifold admits a Kahler metric can be determined by the spectrum of $\square$ (Gilkey [274]) and similarly, whether or not a Kahler manifold is of constant holomorphic sectional curvature can be determined by the spectrum of $\square$ (Gilkey [276]). We conclude this subsection by stating the following important theorem.

THEOREM 4 (Gilkey-Sacks, Donelly)

If M is a compact connected Kahler n-manifold M having the same spectrum as $P^n(\mathbb{C})$, then M is biholomorphically isometric with $P^n(\mathbb{C})$.

For most of the results mentioned in this subsection and for their proofs and for further results in this fascinating field with a classical flavour readers are specially recommended to the following important references:

Minakshisundaram-Pleijel [604], H. Weyl [966], Minakshisundaram [603],

Courant-Hilbert [180], Berger [78], [79], [80], Mckean [589], Hormander [411], Patodi [723], [724], [725], Atiyah-Patodi-Singer [64], Atiyah-Bott-Patodi [56], Sakai [807], Gilkey [272], [273], [274], [275], Gilkey-Sacks [276], Donnelly [203], Seeley [826], Duistermatt-Hormander [213], Atiyah [51].

I-2-8

$P^n(C)$ is not the only compactification of $C^n$: let X be a compact connected complex manifold. X is said to be a holomorphic compactification of $C^n$ is there exists a proper closed complex analytic set A of X such that X-A is biholomorphic with $C^n$. $P^1(C)$ is the only holomorphic compactification of C. $P^n(C)$ is, for $n \geq 2$, a holomorphic compactification of $C^n$ but it is not the only holomorphic compactification of $C^n$. For example $P^2(C)$, $P^1(C) \times P^1(C)$ and the Hirzebruch surfaces are all holomorphic compactifications of $C^2$.

A remarkable theorem of Kodaira [488] is that all the holomorphic compactifications of $C^2$ are rational surfaces; Morrow [626] has given a classification for these compactifications. If we consider the complex surface $C^* \times C^*$ (where $C^*$ is the nonzero complex numbers), the situation is different. There are even non-Kahler compactifications of $C^* \times C^*$. Very recently Ueda [930] has given a classification of all compactifications of $C^* \times C^*$ and Simha [855] has characterized all non-rational structures on $C^* \times C^*$. The situation in higher dimensions must be very difficult indeed. The outstanding problem is to find all the compactifications of $C^n$, $n > 2$. Since $P^n(C)$ has second Betti number 1, a more managable problem is:

PROBLEM 1. Find all the holomorphic compactifications of $C^n$, $n > 2$, with second Betti number equal to one and give a classification of these compactifications.

Similarly, we can ask:

PROBLEM 2. Find all the holomorphic compactifications of ${C^*}^n$, $n > 2$ and give a classification of them.

The first problem was posed by Hirzebruch [735] (then for $n \geq 2$). For results in this direction see Remanujam [765], Morrow [626], Brenton-Morrow [115].

## SECTION 3

### The Problem of Embedding of Complex Manifolds

Part A: The Kodaira Embedding Theorem

Part B: Compact Non-algebraic Manifolds

Part C: Stein Manifolds

### PART A THE KODAIRA EMBEDDING THEOREM

A compact Riemann Surface (that is, a compact connected complex manifold of complex domain one) has non-constant global meromorphic functions. This is a fundamental existence theorem of Riemann. This theorem is not true for higher dimensional compact complex manifolds: there are compact complex manifolds with non-constant global meromorphic functions. (See e.g. Calabi-Rosenlicht [138] and Chapter III of this volume for such examples.)

The above theorem of Riemann leads to another important theorem that any compact Riemann surface X admits an embedding in $P^n(C)$ for some n (that is, X is projective algebraic variety). In fact X is algebraic: it is the zero set of a family of homogeneous polynomials. (For a proof of the embedding theorem, see e.g. Griffiths-Harris [319] or Narasimhan [675] to mention just two places.) Hence projective algebraic varieties are also referred to as algebraic varieties.

Now we ask a basic question: should a given compact connected complex n-manifold, $n \geq 2$, admit an embedding in $P^N(C)$ for some N? If it does, we say it is projective algebraic (or simply algebraic). A fundamental theorem of Chow [161] says that any analytic subvariety of $P^N(C)$ is algebraic, meaning that it is the set of zeros of a finite number of homogeneous polynomials in homogeneous coordinates. The analytic subvarieties of $P^N(C)$ are therefore easier to deal with; in particular, every meromorphic function on an analytic subvariety is rational.

There are 'lots of' (projective) algebraic manifolds and there are 'lots of' non-(projective) algebraic manifolds.

It should be clear from the definition that an algebraic manifold must necessarily be Kahler. But this is not a sufficient condition. It is well known that every complex tori is Kahler but there are complex tori which are not algebraic (see e.g. de la Harpe [182]). Only a restricted class of Kahler manifolds (in Kodaira's terminology:

Kahler varieties of restricted type) turn out to be algebraic.

Consider a compact Kahler manifold with Kahler form w. By de Rham's theorem this determines an element [w] of the de Rham group $H^2(M,C)$. Consider the canonical map: $H^2(M,Z) \to H^2(M,C)$. If [w] is in the image of this map, we say that the Kahler metric is a Hodge metric. This is equivalent to the existence of a positive holomorphic line bundle on M. A compact Kahler manifold is called a Hodge manifold if it admits a Hodge metric (equivalently, if there exists a positive line bundle on it). It is not difficult to prove that al algebraic manifold is Hodge. The celebrated Kodaira embedding theorem is that this condition (of being a Hodge manifold) is also sufficient for being algebraic.

THEOREM 1 (Kodaira Embedding Theorem)

A Hodge manifold is algebraic.

Together with Chow's theorem, this theorem in principle reduces hard analytic problems to simpler algebraic problems. The important step in the proof of the above theorem is another deep result of Kodaira, called the Kodaira vanishing theorem.

THEOREM 2 (Kodaira Vanishing Theorem)

Let X be a compact connected complex manifold with a positive holomorphic line bundle $L \to M$. Let F be any coherent analytic sheaf on X. Then there exists an integer $k_o$ (depending on L and F) such that for any $k > k_o$; $H^q(X, F \otimes L^k) = o$ for every $q \geq 1$.

The above embedding theorem 1 can be considered as a generalization of the theorem that a compact Riemann surface is algebraic. We now re-formulate the above theorem as follows:

THEOREM 1 (Kodaira Embedding Theorem)

A compact complex manifold M is algebraic if and only if there exists a positive holomorphic line bundle of M.

This characterization is very useful when one considers compact complex spaces. Grauert [295] proved that a compact complex space having a positive line bundle is algebraic. For a proof of Kodaira's theorem using Grauert's method see Narasimhan [685].

Some important applications of Kodaira's embedding theorem are:

1) If the Ricci form of a compact complex manifold is positive or negative, then it is algebraic.

2) If a compact Kahler manifold has $H^2(M,\Theta) = 0$, then it is algebraic ($\Theta$ is the sheaf of germs of holomorphic functions on M)

3) If the universal covering manifold of a compact complex manifold M is a bounded domain in $C^n$, then M is algebraic.

4) $\pi: M \to N$ be a branched covering with M and N compact. Then if N is algebraic, M is also algebraic (Grauert).

<u>Moishezon Manifolds</u> Consider the field C(M) of meromorphic functions on a compact connected complex manifold M. A theorem of Siegel [852] and Thimm [896] says that it is a finitely generated extension of C and its transcendence degree cannot exceed the complex dimension of M. Compact connected complex manifolds M of algebraic dimension exactly equal to the dimension of M form an important class, and this class was extensively studied by Moishezon, and these manifolds are now named after him. A compact connected complex manifold of algebraic dimension equal to the dimension of M is called a Moishezon Manifold. A Moishezon manifold is algebraic if and only if it is Kahler [611]. The Kahler condition is necessary as there exist examples of Moishezon manifolds which are not algebraic (e.g. see page 375, Shaferavich [803]). However, every Moishezon surface is algebraic (that is the assumption of Khalerity is unnecessary in this case; see Kodaira [481], Chow-Kodaira [163]). For examples of Moishezon manifolds which are not algebraic, see Shaferavich (page 375, [803], Hartshorne (page 444, [356]).

<u>Moishezon complex spaces</u> For a compact irreducible reduced complex space the theorem of Sirgel-Thimm is also true. A complex space is called a Moishezon complex space if its complex dimension is equal to its algebraic dimension. One can also introduce the notion of a Kahler metric on complex spaces. Recall that if w is Kahler, or associated to a Kahler metric on a complex manifold M, then there exists an open covering $\{U_i\}$ of M and real $C^2$ functions $P_i$ on $U_i$ such that $w = \partial\bar{\partial} P_i$ on $U_i$ and this property in fact characterizes Kahler forms. These functions $P_i$ are strictly pluri-subharmonic in $U_i$ and on $U_i \cap U_j \neq \phi$, $P_i - P_j$ are pluri-harmonic

(that is, they are the real parts of holomorphic functions). Pluri-subharmonic and strictly pluri-subharmonic functions make sense in any complex space. Hence this characterization of Kahler forms on complex manifolds can be extended to complex spaces as well.

Kahler Complex Space  Let M be a complex space (irreducible and reduced). By a Kahler metric on M we mean an open covering $\{U_i\}$ of M together with $\{P_i\}$ where $P_i$ are strictly pluri-subharmonic $C^\infty$ functions such that $P_i - P_j$ is pluri-harmonic in $U_i \cup U_j$. Two such collections $(U_i, P_i)$ and $(V_j, Q_i)$ are said to define the same Kahler metric if each $P_i - Q_j$ is pluri-subharmonic on $U_i \cap V_j \neq \phi$. A complex space is Kahlerian if there exists a Kahler metric on it. Certainly all projective algebraic complex spaces are Kahlerian in this sense. Now consider a Moishezon Kahlerian complex space M. Contrary to the case of Moishezon Kahlerian manifolds, M need notbe projective algebraic. Moishezon [613] has given an example of a Moishezon Kahlerian (singular) surface which is not projective algebraic.

Moishezon manifolds and spaces are 'nearly' algebraic in the sense that they become algebraic after a finite number of blow ups with non-singular centres. More precisely, a Moishezon manifold is biomeromorphic to an algebraic variety. Moishezon complex spaces are also 'algebraic spaces' in the sense of Artin [39].

Topology of algebraic Manifolds  The topology of any compact Riemann surface is well understood. However the topological structures of algebraic manifolds are quite complicated. See, for example, Moishezon [614], Mandelbaum-Moishezon [566]. Even for algebraic surfaces there are serious difficulties with the topological classification. The major difficulty is that given any finite group there exists an algebraic surface whose fundamental group is the given one (Serre). For an account of the topology of algebraic varieties see Chapter VII of Shaferavich [803].

## PART B COMPACT NON-ALGEBRAIC MANIFOLDS

There are many non-Kahler manifolds which are also of interest, e.g. the Hopf manifolds. Recall the theorem of H. Cartan that if G is a property discontinuous subgroup of the group of the holomorphic automorphisms of a complex manifold M, then M/G, the quotient space, has the structure of a complex analytic space, with singularities at the fixed points of G. Let $W = C^n - (0)$. Let G be the infinite cyclic group of holomorphic automorphisms of W generated by g: $(Z_1, \ldots Z_n) \to (\alpha Z_1, \ldots \alpha Z_n)$ where $|\alpha| \neq 1$. It is easy to verify that G acts on W properly discontinuously and fixed point free. Hence $M = W/G$ is a complex manifold. We can show that M is diffeomorphic to $S^1 \times S^3$. These compact complex manifolds (called Hopf manifolds) were first introduced by Hopf [400]. They are non-Kahler since they do not satisfy the necessary condition for Kahlerity that the odd Betti numbers must be even. Kodaira [481], [484] has studied Hopf surfaces in great detail. He has shown that every complex structure on $S^1 \times S^3$ must be a Hopf surface $C^2 - (0)/G$, for a suitable G. It is now known whether every complex structure on $S^1 \times S^{2n-1}$ for $n \geq 3$ is an n-dimensional Hopf structure. Kodaira has given a topological characterization of Hopf surfaces as follows:

THEOREM 3 (Kodaira [481])

Let S be a complex analytic surface such that its second Betti number is zero and its fundamental group contains an infinite cyclic subgroup of finite index. Then S is a Hopf surface.

Generalizing the construction of Hopf, Calabi-Eckmann [137] constructed distinct complex structures on $S^{2p+1} \times S^{2q+1}$, all of which are non-Kahler. The Brieskorn manifolds previously referred to are still more general than the Calabi-Eckmann manifolds.

A number of interesting results have become known in the last few years on complex structures on a product of two odd-dimensional spheres (see e.g. Kato [444], [445], [446], Maeda [561], Morita [623]).

## PART C STEIN MANIFOLDS

Given a noncompact complex manifold, the first natural question to ask is whether it can be imbedded or properly embedded in $C^N$ for some N and, if possible, what is the least N for which this is possible. This question does not arise for compact complex manifolds since there are no nontrivial connected compact submanifolds of $C^N$ (the maximum-modulus principle).

Let us first recall some important facts regarding the corresponding problem for differentiable and real analytic manifolds. A differentiable map $f: X \to Y$ between differentiable manifolds is called an immersion if the rank of f is equal to the dimension of X at all points of X. An immersion is called an imbedding if it is a homeomorphism onto its image in the induced topology. An imbedding is a one-to-one immersion but not every one-to-one immersion is an embedding. A standard problem in differential topology is to find the least integers $\ell$,m such that a differentiable n-manifold X can be immersed in $R^{n+\ell}$ and embedded in $R^{n+m}$.

Quite a number of results are known in this direction. We mention only three well-known important results. Whitney was the first to consider such problems. He proved that a differentiable n-manifold, compact or noncompact, can be differentiably immersed in $R^{2n-1}$ and embedded in $R^{2n}$. The embedding theorems of Morrey [624] and Grauert [292] say that the existence of $C^1$-immersion (embedding) implies the existence of real analytic immersion (embedding). A theorem of Hirsh [371] says that a noncompact differentiable n-manifold can be differentiably immersed in $R^n$ itself if its tangent bundle is trivial.

The complex manifolds, necessarily noncompact, which can be embedded as closed submanifolds in $C^N$ for some N are the Stein manifolds. Just as among compact complex manifolds the algebraic manifolds are the most important, so among the non-compact complex manifolds the Stein manifolds are the most important.

The concept of a Stein manifold is a generalization of the concept of a domain of holomorphy (recall a domain $\Omega$ in $C^n$ is a domain of holomorphy if there exists a holomorphic function in $\Omega$ having every boundary point as a singularity). Stein manifolds and Stein spaces (and further generalizations like q-complete manifolds and spaces) have been extensively developed in the last 25 years. Standard references are Hormander [410], Gunning-Rossi [339], Grauert-Remmert [306]. We now give the definition of a Stein manifold in its original formulation.

DEFINITION

A complex manifold M is called a Stein manifold if: (1) M is holomorphically convex; (2) the global holomorphic functions of M separate points of M; (3) for each point of M there exist global holomorphic functions giving local co-ordinates of that point.

Holomorphic convexity of M means the following. Let K be any compact set of M and let $\hat{K}$ denote the set

$\{x \in M \mid |f(x)| \leq \sup_{x \in K} f(x)$ for any holomorphic function $f$ on $M\}$

$\hat{K}$ is called the holomorphic hull of K. If for each compact K, K is also compact, we say that M is holomorphically convex. Note that a Stein manifold is necessarily non-compact. A domain of holomorphy in $C^n$ is Stein, and in fact the definition of a Stein manifold is given in such a way that it abstracts the properties of domains of holomorphy. With the powerful tools of sheaf theory and cohomology theory, H. Cartan and Serre reformulated the pioneering work of Oka and we have the following celebrated Theorems A and B. Let $\Theta$ be the sheaf of germs of holomorphic functions on a Stein manifold M and let F be any coherent sheaf of $\Theta$-modules (for the notion of coherence refer to Section 1, Part B of this Chapter). Let $H^q(M,F)$ be the cohomology groups with coefficients in F. Then

THEOREM 4 (Oka-Cartan-Serre)

$H^o(M,F)$ generate the stalk $F_x$ for each $x \in M$.

THEOREM 5 (Oka-Cartan-Serre)

$H^q(M,F) = 0$, for all $q \geq 1$.

Moreover theorem B characterizes Stein manifolds. In fact, if $H^1(M,F) = 0$ for any coherent sheaf F of $\Theta$-modules, M must be a Stein manifold. The various nice properties possessed by Stein manifolds are largely due to these theorems A and B. Some of these important properties are: (a) every meromorphic function on a Stein manifold is the quotient of two global holomorphic functions; (b) the additive Cousin problem always has a solution on a Stein manifold; and (c) the multiplicative Cousin problem has a solution on a Stein manifold M if $H^2(M,Z) = 0$ (for these details refer to Chapter V of Hormander [410].

We state now the embedding theorem (Remmert [783], Bishop [87], and Narasimhan [681].

THEOREM 6 (Bishop-Narasimhan-Remmert)

A Stein manifold of dimension n admits a holomorphic closed embedding in $C^{2n+1}$.

Forster [244] improved this result as follows:

THEOREM 7 (Forster)

(a) A Stein manifold of dimension n, $n \geq 2$, can be embedded in $C^{2n}$.

(b) A Stein manifold of dimension n, $n \geq 6$, can be embedded in $C^{2n-k}$ and can be embedded as a closed submanifold in $C^{2n+1-k}$ where $k = [\frac{n-2}{3}]$ ( [ ] denotes the integral part).

Currently researchers are trying to see whether the following could be true.

PROBLEM Can a Stein manifold of dimension n, $n \geq 6$, be embedded as a closed submanifold of $C^{2n+1-[\frac{n}{2}]}$?

Motivated by Hirsh's theorem, we can also ask;

PROBLEM If M is a Stein manifold of dimension n whose tangent bundle is trivial, can M be immersed in $C^n$ itself?

Gunning and Narasimham [340] have proved this result for n = 1 (i.e. for open Riemann surfaces).

We do not want to pass up this opportunity without mentioning two famous problems pertaining to Stein manifolds: the Levi Problem and the Serre Problem.

THE LEVI PROBLEM

Let $\Omega$ be a domain in $C^n$ and let $u: \Omega \to R$ be a $C^2$ function. u is called a pluri-subharmonic function if the hermitian form $\sum_{j,k} \frac{\partial^2 u}{\partial Z_j \partial \bar{Z}_k} a_j \bar{a}_j$ is positive semi-definite at every point of $\Omega$. If this form is positive definite $\Omega$ is called strongly (strictly) pluri-subharmonic.

Let X be a complex manifold and D a relatively compact open set of X. Let $a \in \partial D$. D is said to be pseudo convex at a if there exists a neighbourhood U of a in X and a pluri-subharmonic function u on U such that

$$U \cap D = \{x \in U \mid u(x) < 0\}.$$

If U and u can be so chosen that u is strongly pluri-subharmonic satisfying the above condition, then D is said to be strongly pseudo convex at a.

The following theorem characterizes Stein spaces in terms of certain pluri-subharmonic functions.

THEOREM 8 (Narasimhan [682])

(a) Let S be a complex space. Then S is Stein if and only if there exists a continuous strongly pluri-subharmonic function $u: S \to R$ such that $S_\alpha = \{s \in S \mid u(s) < \alpha\}$ is relatively compact in S for all $\alpha \in \mathbb{R}$.

(b) In a Stein space S any set of the above form $S_\alpha$ given by a continuous pluri-subharmonic function is itself Stein and $(S, S_\alpha)$ is a Runge pair (meaning that every holomorphic function on $S_\alpha$ can be approximated uniformly on compact subsets of $S_\alpha$ by holomorphic functions on S).

Part (a) of the theorem was first proved by Grauert [292] for Stein manifolds.

In the year 1910 E.E. Levi found that a domain of holomorphy $\Omega$ in $C^n$ with a smooth boundary is pseudo convex at every boundary point. He asked whether the convex is true.

THE LEVI PROBLEM Whether any domain in $C^n$ with pseudo convex boundary is a domain of holomorphy.

This was solved in 1954 for the domain $C^n$ for all n by Oka, Norguet and Bremermann.

There are generalizations, analogues and stronger versions of this problem in Complex manifolds and spaces, and for solutions from particular cases to the complete solution of this problem in Complex spaces, and for related problems of current interest, readers are particularly referred to Narasimhan [688], Pflug [730], and Siu [859].

Consider the following generalization of the Levi problem: let $\pi: X \to S$ be a holomorphic map where S is a Stein manifold and X is any complex manifold. Assume that for each $s \in S$, there exists a neighbourhood U of s in S such that $\pi^{-1}(U)$ is a Stein manifold. The problem is whether X must be Stein. Taking $S = C^n$, X a domain in $C^n$ and $\pi$ the natural injection map, we get the original Levi problem. In 1953, Serre [828] highlighted a particular case of the general Levi problem.

THE SERRE PROBLEM Let $\pi \to S$ be a holomorphic locally trivial fibre bundle with both the base S and fibre being Stein manifolds. Is the total space X a Stein manifold? (The assumption of a holomorphic locally trivial fibre bundle means that there exists a covering $S_\alpha$ of S such that $\pi^{-1}(S_\alpha) \approx S_\alpha \times F$ for all $\alpha$ where F is a complex manifold).

In the last 25 years this problem had attracted a lot of attention and a number of particular cases were solved. But very recently counter-examples have been found to the Serre problem. We refer readers again to the three references given above for details and the current status of this and related problems.

In the last few years important results have been proved regarding strongly pseudo convex domains; they have been found to be 'more nice' than bounded domains of holomorphy. We list some of the important results.

THEOREM 9 (C. Fefferman [239])

Any biholomorphic equivalence of two strongly pseudo convex domains with smooth boundaries in $C^n$ must be smooth up to the boundary.

REMARK

The boundary has the structure of what is known as a Cauchy-Riemann manifold. Cauchy-Riemann manifolds are being vigorously investigated and recently their local invariants have been calculated, notably by Chern-Moser [157].

See the references given in Chern-Moser [157] for earlier work. Also see Problem 2 on page 51 in Browder [126] and Chern's remarks following the problem.

THEOREM 10 (Diederich-Fornaess [190])

Let $\Omega$ be a pseudo convex domain in $C^n$ with a smooth boundary and let U be a neighbourhood of a in $C^n$ such that $U \cap \partial \Omega$ contains the germ of a real analytic set whose holomorphic dimension is at least q. Then $U \cap \partial \Omega$ contains the germ of a complex analytic set of dimension at least q.

THEOREM 11 (Diederich-Fornaess [192] - Kohn [500])

Let $\Omega$ be a bounded pseudo convex domain $C^n$ with a smooth real analytic boundary. Then (A) $\overline{\Omega}$ has a fundamental system of pseudo convex neighbourhoods (B) for any $a \in \partial \Omega$ and any $q > 0$, the $\overline{\partial}$ - Neumann problem is subelliptic at a for forms of type $(p,q)$.

For details on the $\overline{\partial}$ - Neumann problem and for the notion of subelliptic estimates see e.g. Greiner-Stein [308].

CURVATURE AND STEIN MANIFOLDS Just as the complex projective spaces can be characterized by curvature properties, so we may ask whether Stein manifolds can be characterized by curvature conditions. Let S be a Stein manifold of dimension n. Then by the Bishop-Narasimhan-Remmert embedding theorem, S can be regarded as a closed submanifold of $C^{2n+1}$. The standard Kahler metric on $C^{2n+1}$ induces a Kahler metric on S and it is a complete metric on S. So we ask should every complete Kahler noncompact manifold be Stein. The answer is no, since there are examples of such manifolds which are not Stein. For example, consider the noncompact manifold $P^2(C) \times C^2$; this is easily checked to be a complete Kahler manifold. This is not Stein since it obviously has compact submanifolds but no Stein manifold can have nontrivial compact submanifolds (Maximum Principle). However, we have the following result of Grauert [292].

THEOREM 12 (Grauert)

If S is a domain in $C^n$ with a smooth real analytic boundary admitting a complete Kahler metric, then it is Stein.

More recently Green-Wu [306] have proved the following important theorem.

THEOREM 13 (Green-Wu)

If S is a connected noncompact complete Kahler manifold whose sectional curvature is everywhere positive, then S is Stein.

There are results available strongly supporting the following conjectures.

CONJECTURE A connected noncompact complete Kahler manifold with positive holomorphic bisectional curvature is Stein.

CONJECTURE A simply connected Complex manifold is Stein if and only if it carries a complete Kahler metric of nonpositive holomorphic sectional curvature.

The specific references for this topic are the papers of Green and Wu [304], [305], [306], [307] and Siu-Yau [860].

## SECTION 4

## THE MODULI PROBLEM

We have seen that the existence of a complex structure on a differentiable manifold $\underline{M}$ is something very special. There may not be any complex structure on $\underline{M}$; there may be a unique complex structure on $\underline{M}$ ; there may be countably or uncountably many distinct complex structures on $\underline{M}$ . To determine all the complex structures on a given $\underline{M}$ is the most difficult problem. We have also seen that there may also be exotic complex structures on $\underline{M}$.

Let $\underline{M}$ be a compact connected oriented differentiable manifold of even dimension. Let $\Sigma(\underline{M})$ be the set of isomorphism classes of complex structures consistent with the given differentiable structure and orientation. 'The problem of Moduli' is the problem of prescribing a 'natural' complex analytic structure on $\Sigma(\underline{M})$ and studying its properties.

There are serious difficulties for introducing a complex structure on $\Sigma(\underline{M})$; even if a complex structure exists on $\Sigma(\underline{M})$ it may turn out to be non-Hausdorff (which happens when jumping of structures occurs. See Chapter II). Examples show that, even in dimension 1, it is too much to expect a complex manifold structure on $\Sigma(\underline{M})$. The best we can hope for is the existence of a complex space (with singularities) structure on $\Sigma(\underline{M})$ and we require this complex structure to be 'natural' in a sense explained in the sequel.

I) The Moduli problem has its origin in Riemann's work. Riemann considered the set of all conformal equivalence classes of compact one-dimensional manifolds of fixed genus $g \geq 2$ and observed that these classes can be considered in some sense a continuum described by $(3g - 3)$ complex parameters. He found that the number of independent parameters on which compact one-dimensional manifolds of genus $g$ depend is $3g - 3$, $1$ or $0$ according as $g > 1$, $g = 1$ or $g = 0$. He called the parameters 'Moduli'. Riemann's ideas were made precise after a long time by Teichmuller [895], Rauch [770], Ahlfors [12], [13] and Bers [82], [83]. We have

THEOREM 1 (Riemann-Teichmuller-Rauch-Ahlfors-Bers)

The set of all isomorphism classes of compact Riemann surfaces of genus $g \geq 2$ carries a 'natural' structure of a complex analytic space of dim $(3g - 3)$.

REMARK

The set of all isomorphism classes of compact Riemann surfaces of genus 1 is $\underline{H}/SL(2,Z)$ where $\underline{H}$ is the upper half plane and hence has a nice complex structure. There is only one (isomorphism class of) compact Riemann surface of genus zero.

We briefly indicate how the complex space structure (moduli space) is constructed on the set of all isomorphism (conformally equivalent) classes of compact Riemann surfaces of genus $g \geq 2$. Fix a Riemann surface $R_o$ of genus $g \geq 2$. Consider for any Riemann surface R of genus $g \geq 2$, the homotopy classes H of orientation preserving homeomorphisms of $R_o$ into R. Consider all such pairs (R,H). Call (R,H) and (R',H') equivalent if the homotopy class $H'H^{-1}$ contains a conformal map. The set $T_g$ consisting of conformal equivalence classes $\langle R,H \rangle$ is called the Teichmuller 'space' with reference to the chosen $R_o$. The first main step in the construction is to show that $T_g$ is a complex manifold of dimension $(3g - 3)$. The Teichmuller manifold $T_g$ has many nice properties. It is shown to be a bounded domain of holomorphy. It carries a Kahler metric whose Ricci curvature, holomorphic sectional curvature and scalar curvature are all negative. $T_g$ is topologically the unit ball in $\mathbb{R}^{6g-6}$. The group $H_g$ of homotopy classes of orientation preserving homeomorphisms of $R_o$ onto itself acts on $T_g$ as a transformation group. Each $\alpha \in H_g$ induces the transformation $\langle R,H \rangle \rightarrow \langle R,H\alpha \rangle$ and this action is properly discontinuous and hence $\mathcal{M}_g = T_g/H_g$ is a complex analytic space of dimension $3g - 3$. $\mathcal{M}_g$ is called the Moduli space and this complex structure is known to be 'natural' in the sense defined in the sequel.

For detailed proofs of the above and more information on the Moduli space $\mathcal{M}_g$ and Teichmuller manifold $T_g$, see Ahlfors [12], [13], Bers [82], [83], [84], Earle-Eells [215], Eells [221], Grothendieck [327], Rauch [770], [771], [772], [773].

Compact Riemann surfaces can be regarded as nonsingular algebraic curves over $\mathbb{C}$ and conversely any nonsingular algebraic curve over $\mathbb{C}$ can be regarded as a compact Riemann surface. The moduli variety $\mathcal{M}_g$ for nonsingular algebraic curves of genus g (over $\mathbb{C}$ and more generally over any algebraically closed field K) exists and the structure of this variety has been extensively investigated, notably by Mumford [640], Deligne-Mumford [187]. Deligne and Mumford have shown that $\mathcal{M}_g$, $g \geq 2$, is an irreducible quasi projective variety of dim $3g - 3$. For more on the moduli variety of curves see Mumford [640].

The main aim in higher dimensions is to get global theorems analogous to the above theorem 1. Even in dim 2, only in some very special cases do we have such global results.

The Moduli problem can be divided naturally into two parts: Local theory and Global theory. We have a reasonably complete local theory due to Kodaira, Spencer and Kuranishi. There are three major approaches to the global Moduli Problem. (1) Mumford's geometric invariant theory, Mumford [633], Popp [749], Seshadri [841]. (2) Artin's Theory of algebraic Stacks, Artin [41], (3) Griffiths' theory of Period Matrix domains; Griffiths [313], [314], Griffiths-Schmid [320], Schmid [817], Deligne [183]. (2) is closely related to (1). Our aim in this section is to describe Mumford's theory.

II) We need the notion of a holomorphic family and related notions of deformation theory to explain what we mean by a 'natural' complex structure on $\Sigma(\underline{M})$. We give a brief account of the deformation theory here and a detailed account is given in Chapter II.

Kodaira-Spencer-Kuranishi Theory of Deformations This theory studies complex structures 'close to' a given one: $\Sigma(\underline{M})$ is investigated in a 'neighbourhood' of the point which represents the given complex structure M on $\underline{M}$. Let S be a connected complex analytic space. By a holomorphic family of compact complex manifolds parametrized by S we mean a complex analytic space M together with a proper, surjective simple holomorphic map $\pi: M \to S$. Simplicity of $\pi$ means, by definition, each point of M has a neighbourhood $\underline{U}$ with the following property: there exists an open set U in $\mathbb{C}^n$ and a biholomorphic isomorphism $\underline{U} \to \pi(\underline{U}) \times U$ such that the diagram

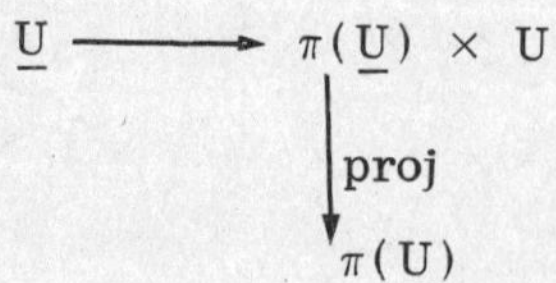


commutes. (A more convenient and useful definition of a family is obtained by requiring the map $\pi$ to be 'flat' instead of simple. Refer to Chapter II). For each $s \in S$, the fibre $\pi^{-1}(s)$, denoted by $M_s$, is a compact complex manifold. We denote the family $(M, S, \pi)$ also by $\{M_s\}_{s \in S}$. It can be proved that the fibres are all diffeomorphic to each other: in fact all have the same real analytic structures. Hence we can regard the complex structures of the fibres to be complex structures on a given compact differentiable manifold $\underline{M}$. Consider a complex structure M on $\underline{M}$. We say a holomorphic family $(M, S, \pi)$ is a holomorphic family of deformations of M if there exists fibre $M_{s_o}$ isomorphic with the given M. We call $s_o$ the reference point and denote the family by $(M, S, \pi, s_o)$. For each tangent vector L to S at $s_o$, 'the infinitesimal deformation of M in the direction of L' is in $H^1(M, \Theta)$, the first cohomology of M with coefficients in the sheaf $\Theta$ of germs of holomorphic vector fields. Hence we have a canonical linear map, the Kodaira-Spencer map $\rho_{s_o}$ : $T_{s_o} S \to H^1(M, \Theta)$ from the Zariski tangent space $T_{s_o} S$ to S at $s_o$ into $H^1(M, \Theta)$. Hence $H^1(M, \Theta)$ controls the local deformation M. Frolicher and Nijenhuis were the first to prove that if $H^1(M, \Theta) = 0$ any family of deformations of M is locally trivial at the reference point: that is, $(M, S, \pi)$ is isomorphic with the product family $(M \times S, S, \rho_s)$ in a neighbourhood of the reference point. When one tries to construct a family of deformations of a given M, one meets with certain 'obstructions'. There is a natural quadratic map: $H^1(M, \Theta) \to H^2(M, \Theta)$ which assigns to every infinitesimal deformation the obstruction to prolong it one step further. If such a prolongation is possible, then one meets another obstruction which also lies in $H^2(M, \Theta)$. If all these obstructions vanish, one can formally construct a family of deformations of M. But then one meets the significant analytic question of convergence. The formal construction need not converge and therefore need not give a genuine family of deformations of M. If $H^2(M, \Theta) = 0$, all obstructions vanish. In this important special case Kodaira-Nirenberg-Spencer proved convergence. That is they proved under the assumption $H^2(M, \Theta) = 0$, there is a family $(M, S, \pi, s_o)$ of deformations of M parametrized by a manifold S such that the Kodaira-Spencer map $\rho_{s_o}$ is an isomorphism at the reference point. If the Kodaira-Spencer map is injective at a point, then we say the family is effectively parametrized at that point.

One of the important notions in the local theory is the notion of an induced family. Let $(M, S, \pi)$ be a family. Let W be a complex space and $f : W \to S$ be a holomorphic map. Then consider $\mathcal{M}_f = \{(x, w) \in M \times W \mid \pi(x) = f(w)\}$. $\mathcal{M}_f$ is a subvariety of $M \times W$. Let $\tilde{\pi} = \pi \times$ (projection to W). Then it can be checked that $(\mathcal{M}_f, W, \tilde{\pi})$

is a holomorphic family and it is called the family induced by f. A family $(M,S,\pi,s_o)$ of deformations of M is said to be versal (or complete) at the reference point $s_o$ if for any other family $(\mathscr{W},W,\sigma,w_o)$ there exists a neighbourhood $W'$ of $w_o$ in W and a holomorphic map $f: W' \to S$ with $f(w_o) = s_o$ such that the restricted family $(\mathscr{W}|W',W',\sigma,w_o)$ is holomorphically isomorphic with the induced family $(\mathscr{M}_f,W',\tilde{\pi},w_o)$. Intuitively a complete family of deformations of M contains all small deformations of M. The fundamental question is whether there exists a complete family of deformations for any compact complex manifold. The celebrated theorem of Kuranishi is that for any given compact complex manifold there exists always a complete family of deformations which is effectively parametrized at the reference point. This is the fundamental theorem of Deformation Theory. A holomorphic family $(M,S,\pi)$ is said to be effectively parametrized if it is effectively parametrized at every point of S. If a family $(M,S,\pi)$ is effectively parametrized at one point $s_o \in S$, it may not be effectively parametrized in a neighbourhood of $s_o$. However if a family is complete at a point, then it is complete in a neighbourhood of that point. For a given compact complex manifold M, we say the number of Moduli $\mu(M)$ is defined if there exists a holomorphic family $(M,S,\pi,s_o)$ of deformations of M which is complete at every point of S and effectively parametrized at every point of S. In this case we define $\mu(M) = \dim S$ and S is called a local space of moduli. Note that $\mu(M)$ need not be defined for a given M. When it is defined $\mu(M) \leq \dim H^1(M,\Theta)$. It follows from the theorem of Kuranishi that when there are no obstructed elements in $H^1(M,\Theta)$, $\mu(M) = \dim H^1(M,\Theta)$. Kodaira-Spencer found many examples, viz., complex tori, hypersurfaces in projective spaces and abelian varieties for which the equality $\mu(M) = \dim H^1(M,\Theta)$ holds. In fact for each of these examples, the effective and complete family is parametrized by a complex manifold. Later Griffiths [312], Kas [441], Mumford [630], Horikawa [403], Burns-Wahl [129] gave important examples of M for which $\mu(M)$ is defined and $\mu(M) \lneqq \dim H^1(M,\Theta)$.

III) Fine and Coarse moduli spaces  Let $\mathbb{D}(S)$ denote the set of equivalence classes $[X]$ of holomorphic families $(X,\pi,S)$ of compact complex n-dimensional manifolds. When S is a point $s_o$, $\mathbb{D}(S)$ is the biholomorphic class of the manifold $X_{s_o}$.

Let (An) denote the category of complex analytic spaces and (Sets) denote the category of sets. Then we have a functor $\mathbb{D}: (An) \to (Sets)$ which associates to

$S \in (An)$ the set $\mathbb{D}(S)$. $\mathbb{D}$ is a contravariant functor and is called the deformation functor. Let $\mathfrak{m}$ be a complex space structure on $\Sigma(\underline{M})$ and $Hol(S,\mathfrak{m})$ denote the set of holomorphic maps from $S \in (An)$ to $\mathfrak{m}$. Then for each $S \in (An)$, we get a well defined map $\alpha(S) : \mathbb{D}(S) \to Hol(S,M)$ by setting $\alpha(S)[X](s)$ to be the bi-holomorphic class of the manifold $X_s = \pi^{-1}(s)$. These maps $\alpha(S)$, $S \in (An)$, define a natural transformation (in the sense of category theory): $\alpha : \mathbb{D} \to Hol(-,\mathfrak{m})$. If $\alpha$ is an isomorphism of functors, we say that the deformation functor $\mathbb{D}$ is representable, $(\mathfrak{m}, \alpha)$ represents $\mathbb{D}$ and $\mathfrak{m}$ is a fine moduli space on $\Sigma(M)$ (or for M). An equivalent definition is that $\mathfrak{m}$ is a fine moduli space if there exists a holomorphic family $(\underline{U}, \pi, \mathfrak{m})$ such that for any holomorphic family $(X, \sigma, S)$, there exists a unique holomorphic map $f: S \to \mathfrak{m}$ with the property that X is isomorphic to the induced family $\underline{U}_f$ (that is $(\underline{U}, \pi, \mathfrak{m})$ has the universal property with respect to holomorphic families).

Because of the possibility of existence of nontrivial automorphisms of fibres, it is realised that it is too much to expect the existence of a fine moduli space. It has been found that more reasonable demands are to require the existence of a complex space $\mathfrak{m}$, a natural transformation (in the sense of category theory) $\alpha: \mathbb{D} \to Hol(-,\mathfrak{m})$ such that $\alpha$ (point) is bijective and $\alpha$ has the universal property with respect to natural transformations For any natural transformation $\beta: \mathbb{D} \to Hol(-,\mathfrak{n})$ there is a unique natural transformation $\gamma : Hol(-,\mathfrak{m}) \to Hol(-,\mathfrak{n})$ such that $\beta = \gamma . \alpha$. If these demands are met, we say $\mathfrak{m}$ (together with $\alpha$) is a coarse moduli space on $\Sigma(\underline{M})$ (or for M). A coarse moduli space, if it exists, is unique in the sense if $(\mathfrak{m}_1, \alpha_1)$, $(\mathfrak{m}_2, \alpha_2)$ are two coarse moduli spaces, then there exists a biholomorphic map $f: \mathfrak{m}_1 \to \mathfrak{m}_2$ such that $\alpha_2 = f \circ \alpha_1$. The main problem is : Does there exist a coarse moduli space for any compact complex n-dimensional manifold ?

It has been found in recent years that for some well known classes of compact complex manifolds coarse moduli spaces exist. Examples of manifolds are known where coarse moduli spaces exist but fine moduli spaces do not. Examples of manifolds for which coarse moduli spaces do not exist are also known. By a natural complex structure for $\Sigma(\underline{M})$ we mean a coarse moduli space for $\Sigma(\underline{M})$.

From the definitions, it is easy to get :

THEOREM 2

A coarse moduli space $(\mathfrak{m}, \alpha)$ is a fine moduli space if and only if there exists a holomorphic family $(\underline{U}, \pi, \mathfrak{m})$ such that

(i) for each $m \in \mathfrak{m}$, $\pi^{-1}(m)$ is in the biholomorphic class of $\alpha(\text{point})^{-1}(m)$

(ii) any two holomorphic families $(X_1, \sigma_1, S)$ and $(X_2, \sigma_2, S)$ are equivalent if and only if the induced maps $f_1: S \to \mathfrak{m}$ and $f_2: S \to \mathfrak{m}$ are the same.

REMARK

The moduli problem can be posed more generally for varieties and schemes. The concepts of fine and coarse moduli spaces can be introduced with necessary modifications. Again fine and coarse moduli are distinct (see Oort [712]). For the moduli problem in these larger categories see Grothendieck [329]; Mumford [633], Seshadri [841], Artin [38], [40], [41], [42], Popp [747], [749], Newstead [695].

## IV Chow Variety, Hilbert Scheme and Douady Space

Let k stand for an algebraically closed field.

1) Chow Variety Recall the definition of an algebraic correspondence (Refer I-1-C-4 (3)). Let $V \subset P^n(k)$ be an r-dimensional projective variety. Consider $(r+1)$ hyperplanes $H_0, \ldots, H_r$ defined by $\sum_j \lambda_{ij} X_j = 0$, $0 \leq i \leq r$. Assume $V \cap (H_o \cap H_1 \cap H_2 \cap \ldots \cap H_r) \neq \phi$. Let $W = P^n(k) \times \ldots\ldots \times P^n(k)$ ((r+1) factors). $(\lambda_{ij})$ can be regarded as coordinates of W. Let $Z = \{(x, \lambda) | x \in v, \Sigma \lambda_{ij} s_j = 0\}$. It can be checked that Z is an irreducible closed subset of $V \times W$. Hence Z is an irreducible algebraic correspondence between V and W. Let $V'$ ($W'$) be the closed image of the projection $Z \to V$ (respectively $Z \to W$). It can be proved that $W'$ is of codimension 1 in W and hence $W'$ is defined by a single equation $F_V(\lambda_{ij}) = 0$. The Form $F_V$ is a homogeneous form of degree d (= degree of V) in each system of variables $(\lambda_{io}, \ldots, \lambda_{iN})$ and is symmetric in the indices i. Moreover the form $F_V$ determines V completely. (See e.g., Safarevich [803], Ch. I, Section 6). $F_V$ is called the associated form of V and its coefficients are called the Chow Coordinates of V. Chow coordinates generalize the classical Plucker coordinates.

Next, instead of hyperplanes take a positive cycle $X = \Sigma N_\alpha V_\alpha$ in $P^N(k)$ of dimension r and degree d $(= \Sigma N_\alpha \deg V_\alpha)$. By the above, each $V_\alpha$ has its associated form $F_\alpha$. The form $F_X = \Pi F_\alpha^{N_\alpha}$ is called the associated form of X. The coefficients of $F_X$ can be naturally regarded as the coordinates of a projective space. The coefficients of $F_X$ (arranged in some fixed order) are called the Chow coordinates of X.

The basic question relevant to our present context is: does the set of all Chow coordinates admit the structure of a variety?

THEOREM 3

Given a projective variety V and integers r and d, the Set of all Chow coordinates of all the positive cycles of dimension r and degree d and contained in V carries a natural structure of a projective variety (called the Chow Variety of V).

For the proof of the theorem (see e.g., Samuel [809], Chapter I, Section 9).

2) Hilbert Scheme The analogue of Chow variety in Schemes is the Hilbert Scheme. The basic question here is: is it possible to parametrize the set of all closed subschemes of a projective space $P^N(k)$, all having the same Hilbert polynomial (for the definition of Hilbert polynomial see the next subsection). A big achievement of Grothendieck is that he constructed in 1960, (see Grothendieck [328]) a scheme parametrizing the set of all closed subschemes, with the same Hilbert polynomial h(n), of a projective space $P^N(k)$.

THEOREM 4 (Grothendieck)

The set of all closed subschemes with the same Hilbert polynomial, of a projective space, carries a natural structure of a scheme.

The parametrizing scheme is called the Hilbert Scheme and is denoted by $Hilb_{N,h(n)}$. Let S denote a locally Noethenian (pre) scheme and let $Hilb_{N,h(n)}(S)$ denote the set of all closed sub (pre) schemes X of $P^N(k) \times S$ such that (i) X is flat and proper over S (there exists a morphism $p: X \to S$ which is proper and flat) and (ii) for $s \in S$, the 'geometric fibre' $S_s$ has h(n) as its Hilbert polynomial. (For the definition of geometric fibre we refer to Part C, Section 1 of Chapter I of this Volume.) Then $Hilb_{N,h(n)}$ is a functor from the category of locally Noethenian (pre) schemes to the category of sets. Grothendieck showed that this functor is representable. The scheme representing $Hilb_{N,h(n)}$ is the Hilbert Scheme of $P^N(k)$ with the given Hilbert polynomial h(n).

An equivalent formulation of the above theorem is that there exists a scheme

$Hilb_{N,h(n)}$ and a family ($\Gamma$, $\gamma$, $Hilb_{N,h(n)}$), universal with respect to flat proper families $p: X \to S$ such that X is a closed subscheme of $P^n(K) \times s$, p is induced by the canonical projection to S, p has a section and the fibres of p all have the same Hilbert polynomial.

The construction of the Hilbert scheme is difficult. For details of the construction see Grothendieck [328]. Grothendieck showed that Hilbert Schemes exist over $\mathbb{Z}$ as well. For moduli theory, Hilbert schemes are of fundamental importance.

3) Douady space Now instead of $P^n(\ )$, consider any complex analytic space W (could be nonreduced) and ask whether the set D(W) of all compact complex subspaces (could be nonreduced) of W can be parametrized by a complex space (may have to be nonreduced). In his thesis Douady [208] proved in 1966 that there exists a natural complex analytic space structure on D(W). We call D(W) with the complex structure constructed by Douady, the Douady space D(W).

THEOREM 5 (Douady)

The set D(W) of all compact complex subspaces of a complex analytic space W carries a natural structure of a (non necessarily reduced) complex analytic space.

This theorem is not only a landmark as such but the techniques used by Douady, especially the theory of Banach analytic spaces, have become standard since then in other contexts as well.

Now assume W to be a complex manifold and consider the set N(W) of all compact complex submanifolds of W. Then Namba [667] has shown that N(W) can be given a complex space structure by a simpler method following Kuranishi [513]. The Namba space N(W) is an open subspace of the Douady's space D(W).

V) Coarse Moduli Spaces for Polarized algebraic manifolds Let M be an algebraic manifold (over $\mathbb{C}$). An ample line bundle on M is said to define a polarization of M. Two ample line bundles define the same polarization if they are numerically equivalent. M together with a polarization L is called a polarized algebraic manifold or simply a polarized manifold and is denoted by (M, L). Two polarized manifolds $(M_1, L_1)$, $(M_2, L_2)$ are said to be isomorphic if there exists an isomorphism $f: M_1 \to M_2$ such that the induced bundle $f^*(L_2)$ on $M_1$ is numerically equivalent to $L_1$.

Two important properties of polarised manifolds are:

(1) The Automorphism group of any polarized manifold is finite (for proof see e.g. Matsusaka-Mumford [583].

(2) Local deformations of a polarized manifold are polarized manifolds (whereas local deformations of an arbitrary algebraic manifold need not be algebraic manifolds (see Section 4, Chapter II of this Volume) (for proof see Matsusaka-Mumford [583]).

A polarized family of polarized manifolds $(X_s, L_s)_{s \in S}$, parametrized by a complex space S is a family $(X, \pi, S)$ together with an ample line bundle $\mathbb{L}$ on X such that $\mathbb{L} \mid X_s$ is numerically equivalent to $L_s$ on $X_s$. As before we have the notion of a polarized family of deformations of a polarized manifold.

For an integer n, let $\chi(M, L^n) = \sum_i (-1)^i \dim H^i(M, L^n)$, be the Euler-Poincare characteristic of (M, L) with coefficients in the sheaf $L^n$ (Note $L^n$ is the tensor product of L with itself n times). It can be shown (see e.g. Hironaka [363]) that for a polarized family $(M_s, L_s)$ parametrized by a connected space S, $\chi(M_s, L_s^n)$ is independent of $s \in S$. $\chi(M, L^n)$ is called the Hilbert polynomial or the Hilbert characteristic function of (M, L). Let us denote it by $h_M(n)$. Let $\Sigma_h$ be the isomorphism classes of polarized algebraic manifolds (M, L) with a fixed Hilbert polynomial. Then we have the following fundamental theorem of Matsusaka [581].

THEOREM 6 (Matsusaka; Matsusaka's Big Theorem)

There exist an integer $m_o$, independent of $(M, L) \in \Sigma_h$, such that m L is very ample for $m \geq m_o$.

REMARKS

1) More explicitly, let h(n) be an integer valued natural polynomial defined on the positive integers. Then Matsusaka proved that there exists a $m_o$ such that for every nonsingular complex projective variety M and every ample line bundle $L \to M$ with $\chi(M, L^m) = h(m)$, then $L^m$ is very ample for $m \geq m_o$.

2) The above theorem is true in characteristic zero. However in arbitrary characteristic the theorem has been proved only for surfaces (see Matsusaka-Mumford [583]).

3) The determination of the exact value of $m_o$ is a very difficult problem. For curves and surfaces of general type $m_o$ has been determined. See Theorem 1, Part 1, Chapter III, and Theorem 14 in Part 3, Chapter III. For $K_3$ surfaces also $m_o$

is known (see Mayer [587], Saint-Donat [804]).

4) The linear system $mL$, $m \geq m_o$, defines a projective embedding $\Phi_m : M \to P^N(C)$, $N = m - 1$. All the manifolds $(M, L) \in \Sigma_h$ are embedded in the same $P^n(C)$ by $\Phi_m$ and it can be shown that all the image varieties $\Phi_m(M)$ have the same Hilbert Polynomial $h_M(mm_o)$.

The Moduli Problem for $\Sigma_h$, an ambitious programme for solving the problem and the difficulties involved in the programme are discussed in detail by Mumford in his famous book [633]. Three good references which elucidate and elaborate the main points of Mumford's book are Seshadri [841], Popp [749], Newstead [695]. Aware of the risk of gross over-simplification, we describe here briefly in broad outline, Mumford's theory, without going into its subtle notions and sophisticated techniques. The programme first leads to the construction of a quotient of a certain locally closed subset of a Hilbert scheme by an equivalence relation defined by the action of an algebraic group with special properties. This then calls for a good theory of quotients in the category of Schemes. Such a theory has been developed by Mumford in characteristic zero. In the category of complex analytic spaces, a nice quotient theory exists for the action of a Lie group (see Cartan [142], Holmann [398], [399]). In the category of algebraic spaces also, existence of quotients for the action of algebraic groups does not pose serious problems (see Popp [747]). However, existence of quotients in the category of schemes in higher characteristic is a far more subtle problem. To tackle this problem Mumford introduced notions of 'good quotients', 'geometric quotients' and 'geometrically reductive algebraic groups' and studied in detail their properties, interrelations and implications. He conjectured that every reductive group is geometrically reductive in any characteristic (the other implication is also true (Nagata)). This is deeply related to the classical invariant theory and Hilbert's fourteenth problem. In 1974 Haboush [341] proved Mumford's conjecture. For a history of this conjecture see e.g. Newstead [695], pages 90 - 92. After this significant development it has become possible to have a satisfactory quotient theory for 'certain subsets' of quasi-projective schemes for the action of geometrically reductive algebraic groups (in any characteristic). The subset alluded to is the set of 'stable points', another key notion in Mumford's theory. The crucial stage of the programme is the extremely difficult analysis of 'stable points' and 'semi-stable points'.

The following are the important results of Mumford:

THEOREM 7 (Mumford)

On the set of all isomorphism classes of polarized abelian varieties with fixed dimension g and degree $d^2$, coarse moduli space exists as a quasi projective algebraic scheme (theorem is true in any characteristic).

THEOREM 8 (Mumford)

On the set of all isomorphism classes of smooth projective curves of genus $g \geq 2$, coarse moduli space exists as a quasi-projective algebraic scheme (this is again true in any characteristic; this theorem is a corollary of the above theorem).

Mumford generalized this theorem to certain types of singular curves called 'stable curves' (for precise definition see Mumford [633]).

THEOREM 9 (Mumford)

On the set of all isomorphism classes of stable curves of genus $g \geq 2$, coarse moduli space exists as a projective variety (this is true in any characteristic).

Mumford considered those varieties of $\Sigma_h$ with what are called 'leven n-structures' (for precise definition see Mumford [633]) and then showed the existence of fine moduli spaces in the special cases of curves, abelian varieties, $K_3$ surfaces, Enriques surfaces and canonically polarized varieties. (For the definitions of various surfaces see Chapter III).

Applying Mumford's theory Giesecker [270], [271] proved the following theorems:

THEOREM 10 (Giesecker)

On the set of all isomorphism classes of surfaces of general type with fixed first Chern number and arithmetic genus, coarse moduli space exists as a quasi-projective variety (this is true only in characteristic zero).

THEOREM 11 (Giesecker)

On the set of all isomorphism classes of stable vector bundles on a smooth projective surface (defined over k) with very ample line bundle, coarse moduli space exists as a projective scheme of finite type over k. (k is an algebraic closed field of any characteristic).

See also Maruyama [571].

Around 1964, Narasimhan and Seshadri [679] took up the programme of constructing moduli spaces for vector bundles of fixed rank r and degree d on a nonsingular irreducible projective curve X of genus $g (g \geq 2)$ over $\mathbb{C}$, by a different method by considering representations of the fundamental group $\pi(X)$. (For classification of vector bundles over the sphere $(g = 0)$ see Grothendieck [323] and over elliptic curves $(g = 1)$ see Atiyah [44]). They showed the existence of moduli space in this case as a nonsingular irreducible quasi-projective variety of dimension $r^2(g-1)+1$. Later Seshadri constructed a natural compactification of this moduli manifold as a normal irreducible projective variety. Narasimhan and Seshadri brought out the connection between their work and Mumford's by showing that the stable bundles of Mumford's theory are precisely those which are given by irreducible unitary representations of the fundamental group. For proofs and more information on their results and for the extension of Narasimhan-Seshadri's theory to arbitrary characteristic, refer to Narasimhan-Ramanan [677], [678], Ramanan [760], Seshadri [839], [842], Narasimhan-Harder [351], Newstead [695].

The existence of coarse moduli spaces for $\Sigma_h$ as algebraic spaces has been established in certain cases by Popp [747].

THEOREM 12 (Popp)

On the set of all isomorphism classes of canonically polarized algebraic manifolds over $\mathbb{C}$ with fixed Hilbert polynomial, coarse moduli space exists as an algebraic space of finite type over $\mathbb{C}$.

In this special case the existence of a coarse moduli space as a complex space is also known and it is due to Narasimhan-Simha [680].

THEOREM 13 (Popp)

Coarse moduli space exists as an algebraic space of finite type over $\mathbb{C}$ in each of the following cases:

1) The set of all isomorphism classes of surfaces of general type with fixed first Chern number and arithmetic genus (compare with Giesecker's theorem, Theorem 10)

2) The set of all isomorphism classes of polarized $K_3$ surfaces with fixed Hilbert polynomial.

3) The set of all isomorphism classes of polarized Enriques surfaces.

Following the method of Narasimhan-Simha [680], we showed in 1975 (unpublished) the existence of a natural complex space structure on the set of all isomorphism classes of compact hyperbolic manifolds. We learnt that around the same time Wright in his thesis had also obtained the same result by essentially the same method. Refer Wright [980].

THEOREM 14 (Wright)

There exists a natural complex space structure on the set of all isomorphism classes of compact hyperbolic manifolds.

Another instance where the moduli problem has been solved is the case of Hopf surfaces. Recall the definition of Hopf manifolds and the important theorem of Kodaira that every complex structure on $S^1 \times S^3$ is a Hopf surface. Kodaira [484] proved

THEOREM 15 (Kodaira)

The set $\Sigma(s^1 \times s^3)$ of all isomorphism classes of Hopf surfaces carries a natural complex analytic structure.

Applications of Moduli theory   Recently it has been found that the moduli theory and its methods can be applied successfully to the study of 'the instantons' of elementary particle physics. Mathematically instantons are certain real algebraic vector bundles over $P^3(\mathbb{C})$. Hence the moduli theory of vector bundles has great relevance to the study of instantons. In the last few years very important results have been obtained. Some important references in this currently active area are Atiyah [55], Atiyah-Ward [65], Atiyah-Hitchin-Drinfold-Manin [61], Atiyah-Hitchin-Singer [62], Singer [856], Narasimhan-Ramadas [676].

Another area where moduli theory has been found applicable is the theory of linear sequential machines. There is a 'geometric invariant theory' of linear machines! See Byrness-Hurt [131] and the references given there.

# 2 Deformation theory of compact complex manifolds

The theory of deformations of compact complex manifolds, due to Kodaira-Spencer-Kuranishi, studies complex structures 'close to' a given one. Let M be a complex structure on a compact connected differentiable manifold $\underline{M}$ and let $\Sigma(\underline{M})$ be the set of all isomorphism classes of complex structures on $\underline{M}$ consistent with the given differentiable structure. Deformation theory studies $\Sigma(\underline{M})$ in a 'neighbourhood' of the point which represents the given complex structure M. The original references for this Chapter are Kodaira-Specer [496], [490] and Kuranishi [512], [513].

## SECTION 1

## THE THEOREM OF KODAIRA-NIRENBERG-SPENCER

The basic notion is that of a holomorphic family.

DEFINITION

A holomorphic family of compact complex manifolds parametrized by a connected complex manifold S is a complex manifold X together with a proper surjective holomorphic map $\pi: X \to S$ of maximal rank.

From the definition it follows that the fibres $\pi^{-1}(s)$, $s \in S$, are all compact complex manifolds of the same dimension. It can be proved that if $f: M \to N$ is a proper surjective holomorphic map between two complex manifolds, there exists a nowhere dense complex analytic set A of M such that f is of maximal rank at any point of M-A and f(A) is a nowhere dense complex analytic subset of N.

DEFINITION

A differentiable family of compact complex manifolds parametrized by a connected differentiable manifold S is a differentiable manifold X together with a map $\pi: X \to S$ which is a proper surjective differentiable map such that: (a) there exists a covering $\{U_i\}$ of X and diffeomorphism $h_i: U_i \to U_i \times V_i$ where $U_i$ is open in $C^n$, $V_i$ is open in S, such that the following diagram commutes:

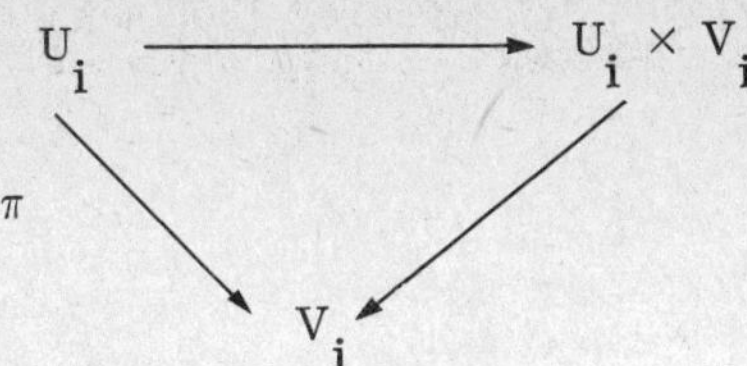


(b) for every s, $s \in S$, $\pi^{-1}(s)$ is a complex manifold and $h_i \mid U_i \cap \pi^{-1}(s)$ is a complex analytic chart of $\pi^{-1}(s)$.

## THEOREM 1

Let $(X, S, \pi)$ be a family of compact complex manifolds (as per any of the definitions above). Then the fibres $\pi^{-1}(s)$ are all diffeomorphic to each other.

## REMARK 1

Hence, in a family the fibres can be regarded as a family of complex structures on a compact differentiable manifold. In general the complex structure of $\pi^{-1}(s)$ depends on s.

## REMARK 2

A continuous family $(X, S, \pi)$ of compact differentiable manifolds parametrized by a connected topological manifold can be defined analogously. However the differentiable structures $\pi^{-1}(s)$ will also be the same.

## DEFINITION

Let $(X, S, \pi)$ be a holomorphic family of compact manifolds. Let M be a given compact complex manifold. We say $(X, S, \pi)$ is a holomorphic family of deformations of M if there exists a point $o \in S$ and a holomorphic isomorphism $\pi^{-1}(o) \approx M$.

## DEFINITION

Let $(X, S, \pi, o)$, $(Y, S, \sigma, o)$ be two holomorphic families of deformations of M. They are equivalent if there exists a neighbourhood $S'$ of o in S such that $\pi^{-1}(s) \cong \sigma^{-1}(s)$ complex analytically for all $s \in S'$. This is an equivalence relation. Hence, we have the notion of the germ of a family of deformations of M.

By deforming M locally we mean to consider the germ of a family of deformations of M.

Also, for studying the local deformations of a compact complex manifold M, it is sufficient to consider germs of families of deformation of M. It becomes necessary to consider holomorphic families parametrized by a complex space which may be reduced or non-reduced.

## Flat Families

> "The concept of flatness is a riddle that comes out of algebra but which is technically the answer to many prayers"
>
> D. Mumford

It has been found, especially by Grothendieck, that a convenient and good definition of holomorphic family for complex spaces and more generally for families of varieties and schemes is obtained by imposing the algebraic condition of flatness on the structure sheaves. This results in the notion of a flat family. We give here some basic facts on flat modules and then give some geometric applications relevant to the present context. Two good references for flat modules are Matsumara [578], Douady [211].

Let A be a commutative ring with unit and let E be a unitary A-module.

DEFINITION

E is said to be flat if for every sequence of A-modules $O \to F' \to F \to F'' \to O$ the sequence $O \to E \otimes F' \to E \otimes F \to E \otimes F'' \to O$ is also exact.

DEFINITION

A free resolution of E is an exact sequence $\dots \to L_n \to L_{n-1} \dots \to L_0 \to E \to O$ where all the $L_i$ are free A-modules.

It is well known that every module has a free resolution. Let F be any other A-module. Then forming the tensor product, we get a sequence

$\dots \to L_n \otimes F \to L_{n-1} \otimes F \to \dots \to L_1 \otimes F \to L_0 \otimes F \to O.$

DEFINITION

We define for $n \geq 1$

$$\mathrm{Tor}_n^A(E,F) = \frac{\mathrm{Ker}(L_n \otimes F \to L_{n-1} \otimes F)}{\mathrm{Im}(L_{n+1} \otimes F \to L_n \otimes F)}$$

and

$$\mathrm{Tor}_0^A(E,F) = E \quad F.$$

The two basic properties of Tor which can be checked easily are that it is independent of the choice of the free resolution and that it is symmetric :

$$\mathrm{Tor}_n^A(E,F) = \mathrm{Tor}_n^A(F,E)$$

The following two theorems give various conditions equivalent to flatness.

THEOREM 2

Let E be an A-module. Then the following conditions are equivalent:

(a) E is flat;

(b) For all A-modules F, and for all $n \geq 1$, $\mathrm{Tor}_n^A(E,F) = 0$;

(c) For all A-modules, $\mathrm{Tor}_1^A(E,F) = 0$.

THEOREM 3

Let A be a Noetherian local ring with maximal ideal m and let $k = A/m$. Let E be any finitely generated A-module. Then the following conditions are equivalent :

(a) E is free;

(b) E is flat ;

(c) $\mathrm{Tor}_1^A(E,k) = 0$.

DEFINITION

Let $\pi : X \to S$ be a holomorphic map between two complex spaces. Then $\pi$ induces a map $\pi^*$ between the local rings $\Theta_{S,s} \to \Theta_{X,x}$; $s = \pi(x)$ by $\pi^*(f) = f \circ \pi$. $\Theta_{X,x}$ can be naturally considered as an $\Theta_{S,s}$-module. We say $\pi$ is flat at x if $\Theta_{X,x}$ is flat $\Theta_{S,s}$-module for $s = \pi(x)$. $\pi$ is flat if it is flat every point of X.

REMARK

1) For examples of flat and non-flat maps see Douady [211] and Hartshorne [356].

2) The above definition of flat holomorphic maps holds beteen non-reduced complex spaces also. However one should be aware of the fact that if $f : X \to Y$ is flat then $f_{red} : X_{red} \to Y_{red}$ need not be flat. (See e.g. Cowsik-Nori [181]). Also in a flat family if a fibre is reduced, neighbouring fibres need not be reduced spaces; if a fibre is irreducible, neighbouring fibres need not be irreducible (see e.g. Hartshorne [356]).

3) The notion of a flat family can be formulated analogously in the case of families of algebraic varieties and of families of schemes.

That flatness is a reasonable condition and has nice properties follows from the following results.

THEOREM 4 (Frish [252], Kiehl [458], Kaup [448]).

(a) For any holomorphic map $\pi: X \to S$ of complex spaces, the set $\{x \in X \mid \pi$ is flat at $x\}$ is open and its complement is a complex analytic set.

(b) If $\pi$ is flat, it is open and conversely if $\pi$ is open, S is nonsingular and the fibres are all reduced complex spaces, then $\pi$ is flat.

(c) If $\pi: X \to Y$ is a holomorphic map between connected complex manifolds, then the following are equivalent (1) $\pi$ is flat (2) $\pi$ is open (3) each fibre of $\pi$ is of pure dimension = dim X - dim Y.

THEOREM 5 (Douady [211])

If S, X are complex spaces and $\pi: S \times X \to S$ be the projection map. Then $\pi$ is flat. In particular, if S,X are complex manifolds and $\pi: X \to S$ is a holomorphic submersion then $\pi$ is flat.

THEOREM 6 (Flatifier theorem, Frish [252])

Given a proper holomorphic map $f: X \to Y$, there exists a flatifier for f. There exists a complex space Z and a holomorphic map $F: Z \to Y$ such that (1) the projection map $p_2 : X \times Z \to Z$ is flat (2) F is universal with respect to the property (1): that is if $G: W \to Y$ is a holomorphic map of complex spaces and if the projection $X \times W \to W$ is flat then there exists a unique holomorphic map $H: W \to Z$ such that $FH = G$. (Notation: $X \underset{Y}{\times} Z = \{(x,z) \in X \times Z \mid f(x) = F(z)\}$).

THEOREM 7 (Sheaf flatifier theorem : Hironaka-Douday)

Let $f: X \to Y$ be a proper holomorphic map and S a coherent $\Theta_X$-module. Then there exists a flatifier for $\Theta_X$: There exists a holomorphic bijective immersion map $F: Z \to Y$ such that (1) $S_Z$ is $f_Z$-flat and (2) F is universal with respect to the property (1).

A theorem of Hironaka [368] asserts that every holomorphic map can be flattened. As a consequence it will follow that any flat bimeromorphic map is biholomorphic. For the definition of a meromorphic map see Section 1 of Chapter III.

THEOREM 8 (Hironaka [368])

Let $f: X \to Y$ be a proper holomorphic map of complex spaces with Y reduced.

Then there exists a proper bimeromorphic map $\pi: Y' \to Y$, a closed imbedding $j: X' \to X \times_Y Y'$ and a flat holomorphic map $f': X' \to Y'$ such that the following diagram commutes

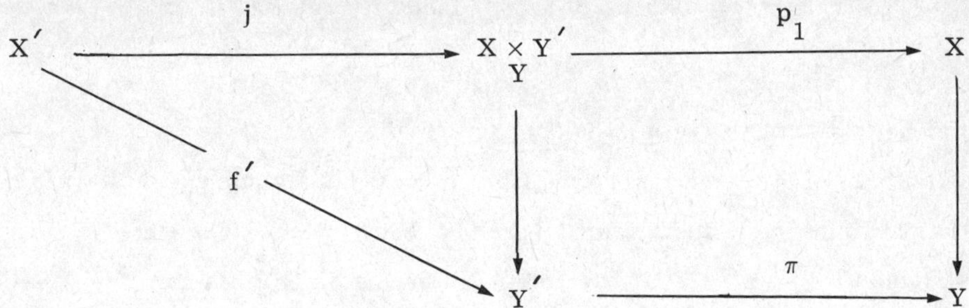


and there exists an open dense subset of U of Y such that j induces a biholomorphic map

$$X \mid (\pi f')^{-1}(U) \approx X \times_Y Y \mid (\pi p_2)^{-1}(U).$$

In a very special case we can give a simple criterion for a map to be flat. Let $\pi: X \to S$ be a finite holomorphic map between complex spaces. Then for each $s \in S$, and for each $x \in X_s \approx \pi^{-1}(s)$, $\Theta_{X_{s,x}} = C \otimes_{\Theta_{S,s}} \Theta_{X,x}$ is a finite dimensional complex vector space.

Let its dimension be $\alpha(x)$. Define $\alpha(s) = \sum_{x \in X_S} \alpha(x)$. This is a finite sum since $\pi$ is finite. Then we have the following criterion for flatness:

THEOREM 9

Let $\pi: X \to S$ be a finite holomorphic map between complex spaces where S is reduced. Then $\pi$ is flat if and only if $\alpha(s)$ is locally constant function of s.

## FLAT HOLOMORPHIC FAMILIES

DEFINITION

Let M be a compact complex manifold. Let $(S, s_o)$ be a germ of a complex space. A deformation of M over S is a complex space X together with a proper flat holomorphic mapping $\pi: X \to S$ and an embedding $M \hookrightarrow X$ which induces an isomorphism $\alpha: M \xrightarrow{\sim} \pi^{-1}(s_o) = M$.

REMARKS

(1) Let $S_1, S_2$ be two representatives of the germ $(S, s_o)$. Let $X_1, X_2$ be two complex equivalent if there exist open neighbourhoods $V_1, V_2$ of $s_o$, $V_1 \subset S_1$, $V_2 \subset S_2$ such that $V_1 = V_2, \pi_1^{-1}(V_1) = \pi_2^{-1}(V_2)$ and $\pi_1 \mid \pi_1^{-1}(V_1) = \pi_2 \mid \pi_2^{-1}(V_1)$. In particular, $\pi_1^{-1}(s_o) = \pi_2^{-1}(s_o) =: \pi^{-1}(s_o)$. Hence a deformation $\pi: X \to S$ of M is an equivalence class of flat proper holomorphic maps.

(2) The flatness of $\pi: X \to S$, when the fibre $\pi^{-1}(s_o)$ is a complex manifold, is equivalent to the following property, for any $x \in \pi^{-1}(s_o)$, $\beta: \alpha^{-1}(U) \times S \to U$ such that $\beta$ restricted to $\alpha^{-1}(U) \times s_o$ is $\alpha$, and the following diagram is commutative:

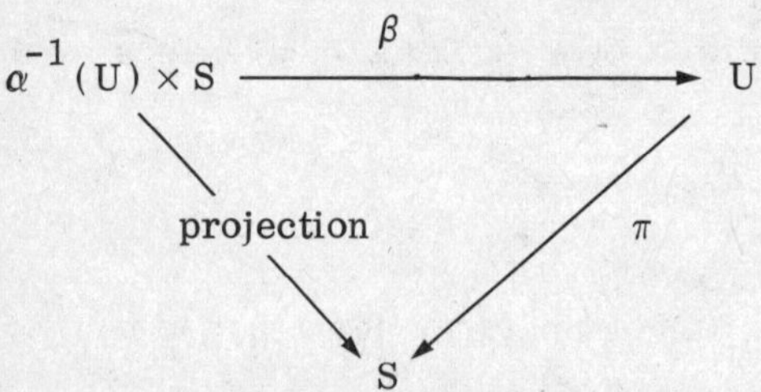


(3) When both X and S are complex manifolds, the above definition is equivalent to Definition given at the beginning of this Section.

(4) The fibre $\pi^{-1}(s_o)$, $s_o \in U$, where U is a small neighbourhood of $s_o$, are all compact complex manifolds and they are all diffeomorphic to each other. The complex structure of $\pi^{-1}(s)$ depends in general on s.

DEFINITION

We say that a compact complex manifold M is rigid if for any holomorphic family $(X, S, \pi, s_o)$ of deformations of M we can find a neighbourhood S of $s_o$ in S such that $\pi^{-1}(s) \approx \pi^{-1}(s_o)$ $(\approx M)$ complex analytically for all $s \in S$.

THEOREM 10

If $H^1(M, \Theta) = 0$, M is rigid (Frolicher-Nijenhuis [258]).

EXAMPLE

(1) $P^n(C)$ is rigid since $H^1(P^n(C) = 0$. This gives strength to the conjecture that $P^n(C)$ is the only one complex structure on $P^n$.

(2) $P^1(C) \times P^1(C)$ is also rigid since $H^1(P^1(C) \times P^1(C), ) = 0$.

Local triviality We have remarked that any family $(X, S, \pi)$ of compact complex manifolds is a differentiable fibre bundle. But it need not be a holomorphic fibre bundle. However, if all the fibres are complex-analytically isomorphic to each other, $(X, S, \pi)$ is a holomorphic fibre bundle according to a theorem of Grauert-Fisher [241]. If $(X, S, \pi)$ is a holomorphic fibre bundle, then in particular it is locally trivial complex analytically. This means that for each $s \in S$ there exists a neighbourhood U of s and a biholomorphic map f of $\pi^{-1}(U)$ onto $U \times M$, such that the following diagram commutes:

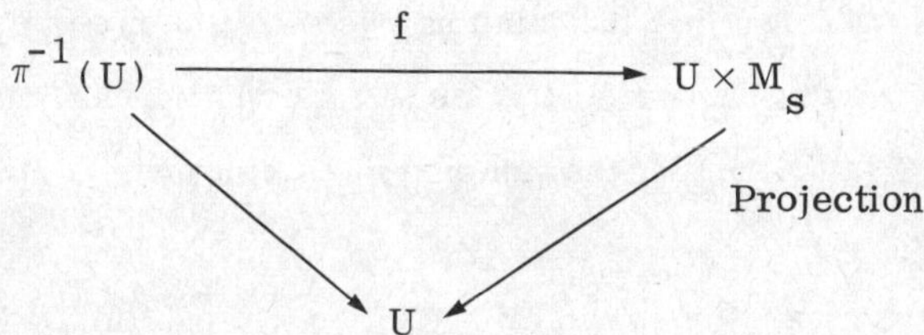


Hence, in a locally trivial family the complex structures of the fibres are complex analytically the same.

Jumping of Complex structures If in a family $(X, S, \pi)$ all the fibres except at one point are biholomorphic to each other, we say there is a jumping of complex structures in the family.

EXAMPLE

(1) Consider the family of Hopf surfaces constructed as follows:

A Hopf surface $M_t$ is defined by

$$M_t = \frac{C^2 - \{0\}}{G_t}$$

where $G_t = \{g^m \mid m \in Z\}$

and

$$g: \begin{bmatrix} Z_1 \\ Z_2 \end{bmatrix} \to \begin{bmatrix} \alpha & t \\ 0 & \alpha \end{bmatrix} \begin{bmatrix} Z_1 \\ Z_2 \end{bmatrix}$$

where $0 < |\alpha| < 1$ and $t \in C$. Then it can be proved that $\{M_t \mid t \in C\}$ is a complex analytic family and $M_t = M_1$ complex analytically for $t \neq 0$ and $M_o \neq M_1$.

(2) Consider the Hirzebruch surfaces $H_n$ in $S^2 \times S^2$ (see I-1-B-6); $H_o = P^1(C) \times P^1(C)$. It can be proved that there exists a holomorphic family

$(X, S, \pi, s_o)$ with dim $S = 1$ such that there exists a neighbourhood $S^1$ of $s_o$ such that for any two integers $n_o, n_1$ with $0 \leq n_1 < n_o$, $\pi^{-1}(s_o) = H_{n_o}$ but $\pi^{-1}(s) \approx H_{n_1}$ for all $s \in S^1 - \{s_o\}$ (see Kodaira-Morrow [489]).

IMPORTANT REMARKS

1) Because of jumping of complex structures, one is unable to generalize the concept of distance between two compact Riemann surfaces. This is the main difficulty in constructing a global moduli space for a compact complex manifold.

2) Grauert [294] has shown that the jumping phenomenon of complex structures of the fibres in a family must happen (if it does) on a complex analytic subset of the parameter space. To this extent, the jumping of complex structures is not that bad a phenomenon to deal with.

3) If jumping occurs at $s_o \in S$ in a family $(X, S, \pi)$, then dim $H^o(X_{s_o}, \Theta_{s_o}) \geq 1$. (See Griffiths [311]).

Kodaira-Spencer map Let $(X, S, \pi)$ be a holomorphic family of compact complex manifolds parametrized by a connected complex manifold S. Then for each $s \in S$ there exists a canonical complex linear map $\rho_s : T_s S \to H^1(M_s, \Theta_s)$ where $T_s S$ represents the complex tangent space of S at s; $M_s$ is the fibre $\pi^{-1}(s)$ over s, and $\Theta_s$ is the sheaf of germs of holomorphic vector fields over $M_s$. This map is called the Kodaira-Spencer map and is of fundamental importance in the study of deformations as it measures the magnitude of the dependence of the complex structure of the fibre $M_s$ on the parameter s. For a tangent vector $v \in T_s S$ the image $\rho_s(v)$ is called the infinitesimal deformation $M_s$ along v. We now outline the construction of $\rho_s$. By definition there exists a covering $\{U_i\}$ of X with the associated biholomorphic maps $h_i : U_i \to U_i \times V_i$. Let $h_i(p) = (Z_i^1, \ldots, Z_i^n, s)$ $s = \pi(p)$: for any $p \in U_i$. 6n $U_i \cap U_k$ we have the transition functions $g_{ik} = h_i \circ h_k^{-1}$:

$$g_{ik} : (Z_k, s) \to (Z_i^1, s), Z_i = g_{ik}^{\alpha}(Z_k, s)$$

where $Z_k = (Z_k, \ldots, Z_k^n)$ and the $g_{ik}^{\alpha}(Z_k, s)$ are differentiable functions of $(Z_k, s)$ in $U_i \cap U_k$ and holomorphic in $Z_k$ for fixed s.

Let $(s^1, \ldots, s^m)$ be local co-ordinates of s on S. For a tangent vector $v = \Sigma v^r (\partial/\partial s^r)$ of S at s we define

$$\Theta_{ik}(s) = \sum_{\alpha=1}^{n} v\, g_{ik}^{\alpha}(z_k, s)\left(\frac{\partial}{\partial z_i^{\alpha}}\right).$$

$\Theta_{ik}(s)$ is a holomorphic vector field on $M_s \cap U_i \cap U_j \cap U_k$. Hence $\{\Theta_{ik}(s)\}$ forms a 1-cocycle of the nerve of the covering $\{M_s \cap U_i\}$ of $M_s$ with coefficients in $\Theta_s$ and hence it determines an element of $H^1(M_s, \Theta_s)$ which we denote by $\rho_s(v)$. We can prove that $\rho_s(v)$ is independent of the choice of the covering $\{U_i\}$ and of the maps $h_i$. Thus we get the map

$$\rho_s : T_s S \to H^1(M_s, \Theta_s)$$

REMARKS

(1) In the case of a differentiable family $(X, S, \pi)$, we can construct in a similar way $\rho_s$ which will be a linear map over the reals.

(2) In the case of a holomorphic family $(X, S, \pi)$ parametrized by a complex space $S, T, S$ is to be interpreted as the Zariski tangent space of $S$ at $s$. In this case, also, the existence of $\rho_s$ can be proved ( see, for example, Forster-Knorr [246 ], Section 5.16 ).

(3) If $\rho_s$ is injective at every point $s$, we say that the family is effective parametrized.

(4) $\rho_s(v)$ gives a measurement of the deformations of the complex structure of $M_s$ along $v$.

Let

$$\Theta_{ik|r}(s) = \sum_{\alpha=1}^{n} \frac{\partial}{\partial s^r}\, g_{ik}^{\alpha}(z_k, s)\left(\frac{\partial}{\partial z_i^{\alpha}}\right).$$

This defines a cohomology class in $H^1(M_s, \Theta_s)$ which we denote by $(\partial M_s / \partial s^r)$.

Define

$$\frac{\partial M_s}{\partial s} = \sum_{r=1} v^r \frac{\partial M_s}{\partial s^r} \quad \text{and} \quad \left.\frac{\partial M_s}{\partial s}\right|_{s=0}$$

to be the cohomology class in $H^1(M_o, \Theta_o)$ given by the cocycle

$$\Theta_{ik} = \sum_{\alpha=1}^{n} \frac{\partial}{\partial s} g_{ik}^{\alpha}(z_k, s)\Big|_{s=o}\left(\frac{\partial}{\partial z_i^{\alpha}}\right).$$

THEOREM 11

If $(X,S,\pi)$ is locally trivial, then $(\partial M_s/\partial s^r) = 0$ for all s and r. Conversely if $\dim H^1(M_s,\Theta_s)$ is independent of s and if $(\partial M_s/\partial s^r) = 0$ for all s and r, the family $(X,S,\pi)$ is locally trivial.

Obstructions If M is a given compact complex manifold, an element $\eta \in H^1(M,\Theta)$ is said to be unobstructed if there exists a holomorphic family $(X,S,\pi,0)$ of deformation of $M \cong M_o$ such that $\eta \in \rho_o(T_o(S)$. If no such family exists, $\eta$ is said to be obstructed. An example of obstruction is given by $M = T^n \times P^1(C)$ where $T^n$ is a complex torus of dim $n \geq 2$. A necessary condition for $\eta$ to be in the image of the Kodaira-Spencer map is that $[\eta,\eta] = 0$. But this is not sufficient. There are higher order obstructions but all these obstructions lie in $H^2(M,\Theta)$. If all these obstructions vanish, we can formally construct a family of deformation of M. In case $H^2(M,\Theta) = 0$, Kodaira-Nirenberg-Spencer proved that we can actually construct a family of deformations of M. Precisely, we have:

THEOREM 12 (Kodaira-Nirenberg-Spencer [490])

Let M be a compact complex manifold with $H^2(M,\Theta) = 0$. Then there exists a holomorphic family $(X,B,\pi)$ where $B = \{t \mid |t| < \in\} \subset C^m$, $m = \dim H^1(M,\Theta)$ such that

(a) $M_o = \pi^{-1}(0) \cong M$.

(b) The map $\rho_o : T_oB \to H^1(M,\Theta)$ is an isomorphism.

REMARKS

(1) If moreover $H^o(M,\Theta) = 0$. the family $(X,B,\pi)$ is effectively parametrized in a neighbourhood of the reference point.

(2) Recently Forster-Knorr [246] have given a new proof of the above theorem by power series methods. See also Forster [245].

(3) As in other existence theorems of deformation theory, the difficult part of the proof of the above theorem is in proving that the formally constructed family of deformations actually converges; the convergence proof of Forster-Knorr is somewhat simpler than the proof given by Kodaira-Nirenberg-Spencer.

NORMAL FAMILIES OF COMPLEX STRUCTURES

Let M now be such that $H^2(M,\Theta) \neq 0$. In Ref. [511] Kuranishi constructed a decreasing filtration of $H^1(M,\Theta) = H^{(1)}$,

$$H^{(1)} \supseteq H^{(2)} \supseteq H^{(3)} \supseteq \ldots \supseteq H^{(n)} = H^{(n+1)} = \ldots ,$$

Denoted by

$$H^*(M) = \bigcap_{n=1}^{\infty} H^{(n)} .$$

Kuranishi proved that there exists a holomorphic family $(X,S,\pi)$ of deformations of $M = M_{s_o}$ which is 'normal' at $s_o \in S$ and such that the map $\rho_{s_o} : T_{s_o} S \to H^2(M,\Theta)$ is an isomorphism onto $H^*(M)$. For details relating to this result see Kuranishi [511], Sundararaman [878], [873].

## SECTION 2

## LOCALLY COMPLETE FAMILIES AND THE THEOREM OF KURANISHI

DEFINITION (induced family)

Let $(X,S,\pi)$ be a holomorphic family of compact complex manifolds. Let W be a complex space and $f: W \to S$ be a holomorphic map. Then f induces a unique family over W as follows. Define $X_W = \{(w,x) \in W \times X \mid f(w) = \pi(x)\}$. This is an analytic set in $W \times X$. Define $\tilde{\pi}: X_W \to W$ by taking $\tilde{\pi}$ = (projection to W) $\times\ \pi$. Then it can be checked that $(X_W, W, \tilde{\pi})$ forms a holomorphic family of deformations. The induced family is also denoted by $(X_f, W, \tilde{\pi})$.

DEFINITION (locally complete family)

Let $(X,S,\pi)$ be a holomorphic family of deformations. It is said to be locally complete at $s_o \in S$ (or versal at $s_o$) if for any other holomorphic family $(Y,W,\rho)$ of deformations of M with reference point $w_o$ there exists a neighbourhood $W'$ of $w_o$ in W and a holomorphic map $f: W' \to S$ with $f(w_o) = s_o$ such that the restricted family $Y \mid W'$ is holomorphically isomorphic with the induced family $(X_{W'}, W', \tilde{\pi})$.

REMARKS

(1) It can be proved that if a family is complete at one point $s_o$, it is complete in a neighbourhood of $s_o$.

(2) We could have formulated the definition of versality to families parametrized by complex manifolds. However, examples show that if a compact complex manifold M

that there are obstructed elements in $H^1(M,\Theta)$, then in any versal family of M the parameter space must have a singularity at the reference point. This is the main reason why one has to consider families parametrized by complex spaces.

(3) Suppose in a holomorphic family $(X,S,\pi)$ parametrized by a complex manifold S is such that at $s_o \in S$, the Kodaira-Spencer map $\rho_{s_o}$ is surjective. Then a theorem of Kodaira-Spencer [495] states that the family must be holomorphically complete at $s_o$. Hence the family of complex tori that we construct is complete at every point of the parameter space.

DEFINITION (local space of moduli)

Let $(X,S,\pi,s_o)$ be a locally complete family of deformations of M, as in the above definition. If the germ of f is unique, i.e. if $M_f$ and $M_g$ are isomorphic, then $f = g$, and we say that the family, $(X,S,\pi,s_o)$ is a universal family or a modular family for M and $(S,s_o)$ is called a local space of moduli for M.

THEOREM 13 (Fundamental theorem of deformation theory - Kuranishi)

For any compact complex manifold M, there exists a versal family of deformations.

More precisely, Kuranishi proved,

THEOREM 13'

Let $A^1$ be the space of $(0,1)$ forms with values in the holomorphic tangent bundle TM. Then there exists a neighbourhood W of the origin in $H^1(M,\Theta)$ and a holomorphic map $\Phi: W \to A^1$ such that if K is the analytic set in W defined by

$$K = \{t \in W \mid IP[\Phi(t), \Phi(t)] = 0\},$$

then the family $\{\Phi(t)\}_{t \in K}$ represents a locally complete family $(X,K,\pi)$ of deformation of M at the reference point 0, and the Kodaira-Spencer map $\rho_o: T_oK \to H^1(M,\Theta)$ is an isomorphism. (IP represents the harmonic projection operator $A^2 \to H^2$).

REMARKS

(1) The versal family $(X,K,\pi)$ constructed by Kuranishi is called the Kuranishi family for M and the parameter space K is called the Kuranishi space of M.

(2) If $H^o(M,\Theta) = 0$, the Kuranishi family is in fact a modular family and the Kuranishi space is a local space of moduli. More generally, if the Kuranishi family

$(X, K, \pi)$, $\dim H^o(M_t, \Theta_t)$ is independent of t, and if K is reduced, then it can be proved that the Kuranishi family is a modular family and the Kuranishi space is a local space of moduli, (see e.g. Wavrik [951]).

(3) It is not necessary that $H^o(M,\Theta)$ should be zero for M to have a local space of moduli, e.g. M = a complex torus.

(4) There are examples of complex manifolds not having a space of moduli, e.g. M = Hirzebruch Surface.

(5) If a local space of moduli exists for a compact complex manifold, it must be the Kuranishi space of that manifold (see e.g. Wavrik [951]).

(6) Mumford [630] and Griffiths [312] gave examples of complex manifolds for each of which the Kuranishi space is nowhere reduced. Also see Horiwaka [405].

(7) If $H^1(M,\Theta) = 0$, the Kuranishi space K reduces to a point so that M is locally rigid.

(8) If $H^2(M,\Theta) = 0$, K is a complex manifold and $\dim K = \dim H^1(M,\Theta)$. The Kuranishi family reduces to the family constructed by Kodaira-Nirenberg-Spencer.

(9) If M is a compact Riemann surface of genus g, the Kuranishi space has dimension equal to the dimension of $H^1(M,\Theta)$, which is $3g - 3$ if $g \geq 2$; 1 if $g = 1$ and 0 if $g = 0$.

(10) The Kuranishi family is effectively parametrized at the reference point but in general it need not be effectively parametrized in any neighbourhood of the reference point even if $H^o(M,\Theta) = 0$.

(11) If $H^1(M,\Theta)$ contains obstructed elements, the Kuranishi space of M must have a singularity at the reference point.

(12) The difficult part of the proof of Kuranishi's theorem is in proving that the family $\{\Phi(t)\}_{t \in K}$ represents a locally complete family. Douady [207] has given a simpler proof.

Regarding the fundamental importance of Kuranishi's theorem, it suffices to quote Chern: "The crowning achievement is the theorem of Kuranishi, which says that for any compact complex manifold, there exists a universal family of deformations"; (Chern [154], page 190; by 'universal' here is meant only versal in the sense we have defined).

## SECTION 3

## OBSTRUCTIONS TO THE EXISTENCE OF A FAMILY OF DEFORMATIONS AND TO THE EXISTENCE OF A LOCAL SPACE OF MODULI

(A) Obstructions to the existence of deformations Let $X_o$ be a compact complex manifold and $\Theta_o$ be the sheaf of germs of holomorphic vector fields of $X_o$. $\Theta_o$ is a sheaf of Lie algebras: there is the bracket operator $[,]: \Theta_o \times \Theta_o \to \Theta_o$. This induces a bracket operator also denoted by $[\,,\,]$, on the cohomology algebra $H^*(X_o, \Theta_o)$. If $a \in H^p(X_o, \Theta_o)$, $b \in H^q(X_o, \Theta_o)$ then $[a,b] \in H^{p+q}(X_o, \Theta_o)$ and $[a,b] = (-1)^{pq+1}[b,a]$. In particular if $a, b \in H^1(X_o, \Theta_o)$ are represented by the cocycles $\{\Theta_{ij}\}$ and $\{\varphi_{ij}\}$, then $[a,b] \in H^2(X_o, \Theta_o)$ is represented by the cocycle $[a,b]_{ijk} = \frac{1}{2}\{[\Theta_{ij}, \varphi_{jk}] + [\varphi_{ij}, \Theta_{jk}]\}$.

Let $(X, S, \pi, s_o)$ be a differentiable family of deformations of $X_o = \pi^{-1}(s_o)$. Consider the Kodaira-Spencer map $\rho_{s_o}: T_{s_o}S \to H^1(X_o, \Theta_o)$. If $u, v \in T_{s_o}S$, it can be checked that $[\rho_{s_o}(u), \rho_{s_o}(v)] = 0$. In particular if $a \in \rho_{s_o}(T_{s_o}S)$, $[a,a] = 0$. We call a class $a \in H^1(X_o, \Theta_o)$ a deformation class if $[a,a] = 0$. $[a,a] = 0$ is a necessary condition for a to be in the image of the tangent space at the reference point of the parameter space of a family of deformations of $X_o$. This is only a primary obstruction; there are higher order obstructions. The sequence of obstructions $w_k$ are defined as follows: $w_1: H^1(X_o, \Theta_o) \to H^2(X_o, \Theta_o)$ is defined by $w_1(a) = [a,a]$. Assuming $w_k$ is defined, $w_{k+1}$ is defined on the subset of $H^1(X_o, \Theta_o)$ where $w_k$ is zero with values in $H^2(X_o, \Theta_o)$ (to be more precise in 'a quotient of' $H^2(X_o, \Theta_o)$ see Douady [206] for details and the following paragraphs). If all these obstructions vanish, then we can formally construct a family of deformations of the given $X_o$. If $H^2(X_o, \Theta_o) = 0$, all these obstructions vanish. Kodaira-Nirenberg-Spender theorem is that in this case there is a genuine family of deformations of $X_o$ parametrized by a complex manifold and this family is locally complete. Kuranishi's theorem is that even when $H^2(X_o, \Theta_o) \neq 0$, it is possible to pass through the obstructions and construct a locally complete family of deformations of $X_o$ but parametrized by a complex space.

A class a $H^1(X_o, \Theta_o)$ is said to be obstructed if $[a,a] \in H^2(X_o, \Theta_o)$ is not zero.

Kodaira-Spencer [496] showed that for the manifold $X_o = P^1(C) \times T^n(C)$; $T^n(C)$ is a complex n-torus, $n \geq 2$, has obstructed classes in $H^1(X_o, \Theta_o)$. See also Sundararaman [879]. Later Kas [441] gave an example of a complex surface $X_o$ with obstructed classes in $H^1(X_o, \Theta_o)$.

(B) Obstructions to the existence of local space of Moduli Let $X_o$ be a compact complex manifold, S be a complex analytic space and let $s_o \in S$. For each open set U of $X_o$ consider open sets W and W of $S \times X_o$ containing $s_o \times U$ and biholomorphic maps $f: W \to W$ such that $p_1 \circ f = p_1$ (where $p_1$ represents the projection map to S) and f is identity on $s_o \times U$. Call two such $f_1, f_2$ equivalent if they agree in a neighbourhood of $s_o \times U$. Denote by $\Gamma(U)$ the set of equivalence classes of such biholomorphic maps. $\Gamma(U)$ is a group under composition. The association $U \to \Gamma(U)$ is a presheaf and the sheaf associated to this presheaf is denoted by $\Gamma$. Note that $\Gamma$ is a sheaf of nonabelian groups. It is called the sheaf of germs of vertical automorphisms. It can be shown that there is a one-to-one correspondence between the cohomology set $H^1(X_o, \Gamma)$ and the set of germs of deformations of X over $(S, s_o)$.

Let $(X, S, \pi, s_o)$ be a family of deformations of $X_o$. The n-th infinitesimal neighbourhood $S^{(n)}$ of $S_o$ in S is ringed space $(s_o, \Theta_S/m^{n+1})$ where m denotes the sheaf of germs of holomorphic functions on S which vanish at $s_o$. Note that the underlying topological space of $S^{(n)}$ is just the point $s_o$. Now restrict the given family to $S^{(n)}$ and denote the restriction by $(X^{(n)}, S^{(n)}, \pi^{(n)})$. Note $X^{(n)} = X_o, \Theta_{X_o}/m^{n+1}\Theta_{X_o})$. By what we said earlier, the set of germs of deformations of $X_o$ over $(S^{(n)}, s_o)$ is in one-to-one correspondence with the cohomology set $H^1(X_o, F^{(n)})$ where $F^{(n)}$ is the sheaf on $X_o$ whose sections over an open U are holomorphic automorphisms of $S^{(n)} \times U$ which are identity on $s_o \times U$. The sheaves $F^{(n)}$ give an exact sequence

$$0 \to K^{(n)} \to F^{(n)} \to F^{(n-1)} \to 1.$$

The kernel sheaf $K^{(n)}$ can be shown to be a sheaf of abelian groups in $F^{(n)}$. Grothendieck [329] has proved that a family of deformations over $(S^{(n-1)}, s_o)$ can be extended over $(S^{(n)}, s_o)$ if and only if its obstruction in $H^2(X_o, \Theta_o) \times m^n/m^{n+1}$ vanishes and that two extensions differ by an element of $H^1(X_o, \Theta_o) \times (m^n/m^{n+1})$.

Suppose $X_o$ has a local space of moduli $(S, s_o)$ and let $(K, 0)$ be the Kuranishi space of $X_o$. It is clear from the definition of S and construction of K, that $\dim T_{s_o} S = \text{Dim } T_o K = \dim H^1(X_o, \Theta_o)$. Since both the parametrizing spaces S, K are locally complete there exist holomorphic maps $\alpha: (S, s_o) \to (K, o)$ $\beta: (K, o) \to (S, s_o)$. Since S is a space of moduli $\beta \circ \alpha = 1$. From this it follows that $\beta$ is a biholomorphic map (Wavrik [951]). Hence if a space of moduli exists for $X_o$, then it must be the Kuranishi space of $X_o$. Hence to find the obstructions to the existence of a local space of moduli; we have to find the obstructions for the Kuranishi space to be the local space of moduli.

Let $(X, K, \pi, o)$ be the Kuranishi family and let $f, g: (S, o) \to (K, o)$ be germs of holomorphic maps such that the induced families $(X_f, S, \tilde{\pi}_1, o)$ and $(X_g, S, \tilde{\pi}_2, o)$ are isomorphic. If $f = g$, it follows that any holomorphic automorphism of the family $X_g^{(n-1)}$ which identity on $X_o$ should be extendable to the family $X_g^{(n)}$ for all n and $f^{(n)} = g^{(n)}$ for all n. The converse is also true but it is more difficult to prove this. We have the following

<u>THEOREM 14 (Grothendieck [329], Wavrik [951])</u>

Let $(X, K, \pi, o)$ be the Kuranishi family for $X_o$. Let $f, g: (S, o) \to (K, o)$ be germs of holomorphic maps such that the induced germs of holomorphic families of deformations on $X_o$ are holomorphically isomorphic. If any holomorphic automorphism of $X_g^{(n+1)}$ which is identity on $X_o$ is extendable to a holomorphic automorphism of $X_g^{(n)}$ which is identity on $X_o$, then $f = g$ and hence the Kuranishi family $(X, K, \pi, o)$ is a local space of moduli for $X_o$.

The obstructions to the extendability of holomorphic automorphisms of $X_g^{(n-1)}$ to $X_g^{(n)}$ are known (see Warvik [951], Douady [206], Forster [245]).

(C) <u>Deformation Spaces</u> Let $(X, S, \pi, s_o)$ be a differentiable family of deformations of a given compact complex manifold $X_o$. The image $\rho_o(T_{s_o} S) \subset H^1(X_o, \Theta_o)$ is called the space of infinitesimal deformations for $X_o$. This space is determined by the particular imbedding of $X_o$ as a fibre in the family $(X, S, \pi)$. Each imbedding of $X_o$ as a fibre in differentiable family $(X, S, \pi)$ determines a subspace of $H^1(X_o, \Theta_o)$, namely the corresponding space of infinitesimal deformations for $X_o$.

A space of infinitesimal deformations is called maximal if it is not a proper subspace of a space of infinitesimal deformations determined by another imbedding of $X_o$ in a a differentiable family. A maximal space D of infinitesimal deformations D is called a deformation space. It follows from this definition that a deformation space D is the space of infinitesimal deformations determined by the imbedding if $X_o$ in a family $(X, S, \pi)$, as a fibre over a point $s_o \in S$; in fact this family $(X, S, \pi)$ is effective at the reference point. It is easy to check that any deformation space D is an abelian Lie algebra (that is $[a, b] = 0$ for any $a, b \in D$.

There are analogues in classical algebraic geometry corresponding to the deformation concepts introduced by Kodaira-Spencer. The following important analogues should be noted.

1) 'continuous systems' correspond to 'differentiable families of deformations'

2) 'characteristic linear systems' correspond to 'spaces of infinitesimal deformations'

3) 'complete continuous systems' corresponds to 'locally complete differentiable families of deformations'.

4) 'characteristic linear systems of a complete continuous system' corresponds to 'deformation spaces'

5) The analogue of the statement "The characteristic linear system of a complete continuous system is complete" is the statement that "the image of the tangent space (at the reference point of a differentiable family of deformations of a compact complex manifold $X_o$) by the Kodaira-Spencer map is all of $H^1(X_o, \Theta)$" (i.e. $\rho_{s_o}(T_{s_o} S) = H^1(X_o, \Theta_o)$).

For further information on this interesting topic see Zariski [993], [997], Kodaira [479], [482], [483], Appendix to Chapter V by Mumford, of Zariski [997], Kodaira-Spencer [498], Mumford [635], Kleiman [466].

## SECTION 4

## STABILITY OF STRUCTURES

It is always interesting and useful to know whether a certain important property possessed by a fibre in a family is enjoyed by neighbouring fibres. The important properties of being Kahler and being Hyperbolic are locally preserved, that of being algebraic is not locally preserved and that of being homogeneous is not locally preserved.

(A) Kahlerity is locally preserved:

Before stating the result, we have to state another deep theorem of deformation theory. The occurrence of the jumping phenomenon in a family $(X, S, \pi)$ depends on the function $d_q : S \to Z^+$ given by $d_q(s) = \dim_C H^q(X_o, \Theta_o)$; jumping occurs when $d_q(S)$ is not locally constant. The theorem of Kodaira-Spencer [496] is that this function is in general an upper semi-continuous function.

THEOREM 15 (The upper semi-continuity theorem)

Let $(X, S, \pi)$ be a holomorphic family with S a complex manifold. Then the function $d_q : S \to Z^+$ given by $d_q(s) = \dim_C H^q(X_s, \Theta_s)$ is upper semi-continuous for each q (that is( $\dim H^q_C(X_s, \Theta_s) \leq \dim H^q_C(X_{s_o}, \Theta_{s_o})$ if s is in a sufficiently small neighbourhood of s, for each q). The proof given by Kodaira- Spencer uses important results of elliptic partial differential equations and harmonic theory (potential theory on compact complex manifolds). As an application of this theorem Kodaira-Spencer proved the following stability theorem.

THEOREM 16 (The local stability of Kahler metrics)

Let $(X, S, \pi)$ be a differentiable family of compact complex manifolds parametrized by a differentiable manifold S. If for some $s_o \in S$, $X_{s_o} = \pi^{-1}(s_o)$ carries a Kahler metric, then for sufficiently small neighbourhood U of $s_o$ in S, all the fibres $X_s$, $s \in U$ admit Kahler metrics. Moreover given a Kahler metric $\sigma_o$ on $X_{s_o}$, we can choose a neighbourhood U of $s_o$ and a Kahler metric $\sigma_o$ on each fibre $X_s, s \in U$, such that $\sigma_o$ depends differentiably on s and $\sigma_{s_o} = \sigma_s$.

This stability theorem holds only locally. Originally Kodaira-Spencer asked the question whether any deformation of a Kahler manifold is Kahler? Hironaka [364] gave an example of a compact Kahler manifold of dimension 3 having non-Kahler deformations. The following problems remain unsettled.

Problem 1: Is any deformation of a Kahler surface Kahler?

Problem 2: Is any deformation of $P^n(C)$, $n > 2$, $P^n(C)$? (Known: any deformation of $P^2(C)$ is $P^2(C)$, Kodaira-Spencer [496], II, Theorem 20.1)

With regard to Problem 1, it should be mentioned that Moishezon [613] has recently shown that there exist singular Kahler surfaces, having non-Kahler local deformations

(Recall the notion of Kahler complex spaces: see section 1, part B, Chapter I).

Grauert's Direct Image Theorem

In his celebrated paper Grauert [293] has generalized the upper semicontinuity theorem to families parametrized by complex spaces. Since methods of potential theory are not available for complex spaces he had to develop entirely different tools. Among other things he used important results of Stein spaces and obtained a theorem (whose proof is very long and tedious), known since then as the Grauert Direct Image theorem. This is considered to be an important landmark. Given a holomorphic map $f: X \rightarrow Y$ between complex spaces and a sheaf $F$ on $X$, associate to each open set $U$ of $Y$, the group $H^q(f^{-1}(U), F)$. This defines a presheaf on $Y$ and the associated sheaf on $Y$ is called the $q^{th}$ direct image of $F$ and is denoted by $(R^q f)F$ (or $f_q(F)$). Grauert's direct image theorem is

THEOREM 17 (Grauert's Direct Image Theorem)

Let $f: X \rightarrow Y$ be a proper holomorphic map between complex spaces and let $F$ be a coherent sheaf on $X$. Then all the direct image sheaves $(R^q f)F$ are also coherent sheaves on $Y$.

Grauert has shown that the Kodaira-Spencer upper semicontinuity theorem follows from this. Moreover Grauert's theorem gives the additional information that the functions $d_q(s)$ are actually locally constant outside of a complex analytic set. Hence in a holomorphic family if jumping of structures occurs it must do so only on a complex analytic set. This additional information could not be got just from the original Kodaira-Spencer upper semicontinuity theorem. There are now simpler proofs available for Grauert's theorem. See e.g. Knorr [468], Narasimhan [686], Verdier-Kiehl [459].

(B) Algebricity is not locally preserved:

Let $(X, S, \pi)$ be a family. Assume $M_{s_o} = \pi^{-1}(s_o)$ is an algebraic manifold. Then in general we cannot expect $M_s$ to be algebraic in some neighbourhood of $s_o$ as the family of complex tori shows.

Let $T^n$ be an n-dimensional complex torus. It is defined by $2^n$ vectors $\omega_1, \ldots \omega_{2n}$ in $C^n$ which are linearly independent over the reals. Let $\omega_1, \ldots, \omega_{1n}$ be the components of the vector $\omega_i$. Form the period matrix $\Omega = (\omega_{ij})$, $1 \leq i \leq 2n$, $1 \leq j \leq n$. Suppose there exists a non-singular skew symmetric matrix $A$ of rank $2n$ such that

(1) $\Omega A^t \Omega = 0$

(2) $-\sqrt{(-1)}\,\Omega A^t \overline{\Omega} = M > 0$ where $A = Q^{-1}$.

Then $\Omega$ is called a Riemannian matrix. An algebraic torus is called an abelian variety. Not every torus is an abelian variety. The following theorem is well known.

THEOREM 18

An n-dimensional complex torus is an abelian variety if and only if any one of the following three equivalent conditions hold (and hence all hold):

(a) The periods defining the torus give rise to a Riemannien matrix;

(b) There exists a positive line bundle on the torus;

(c) The torus is a Moishezon manifold.

It can be proved that, in a family $(X, S, \pi)$ if any one fibre is a torus then every fibre is a torus (see Andreotti and Stoll [30]. p] 339).

We construct a family of complex tori as follows. Let S be the space of $n \times n$ matrices $s = (s^{\alpha}_{\beta})$ with $\det \operatorname{Im}(s) > 0$; from the $n \times 2n$ matrix $\omega(s) = (I, s)$ where I is the identity $n \times n$ matrix. Let G be the group of analytic automorphisms of $C^n \times S$ generated by $g_j : (Z, s) \to (Z + \omega_j(s)), s)$; $j = 1, \ldots, 2n$ where $\omega_j(s)$ is the $j^{th}$ column vector of $\omega(s)$. G acts properly discontinuously without fixed points and hence $X = C^n \times S/G$ is a complex manifold. The canonical projection $C^n \times S \to S$ induces a proper surjective holomorphic map $\pi : X \to S$ and each fibre $\pi^{-1}(s) = M_s$ is an n-torus with periods $\omega_1(s), \ldots, \omega_{2n}(s)$. This family $(X, S, \pi)$ is effectively parametrized and in fact at every point $s \in S$, the Kodaira-Spencer map $\rho_s$ is an isomorphism. Suppose $M_{s_o}$ is an abelian variety; then in every neighbourhood of $s_o$ there exists a point s such that the period matrix of $M_s$ is not a Riemannian matrix and hence $M_s$ is not algebraic.

The following theorems are of interest in this connection.

THEOREM 19 (Moishezon-Tjurin [802])

Let $(X, S, \pi)$ be a holomorphic family and let $M_s = \pi^{-1}(s_o)$ be an algebraic manifold.

(1) Then in some neighbourhood U of $s_o$ in S there exists an analytic set B, with co-dimension not greater than the dimension of the space of two-dimensional holomorphic

forms on $M_{s_o}$, such that for all $s \in B$, $M_s$ is algebraic. The set of points A in S, such that $M_s$ is algebraic for every $s \in A$, is contained in the union of not more than a countable number of such analytic sets.

(2) Let $M_o$ be a Hodge manifold with $H^2(M,\Theta) = 0$; then in any holomorphic family $(X,S,\pi,s_o)$ of deformations of $M_o$ there exists a neighbourhood U of $s_o$ in S such that $M_s$ is Hodge for $s \in U$.

(3) There exists a locally complete family $(X,S,\pi)$ of $K_3$-surfaces such that $\dim s = 20$ and the set $\{ s \in S \mid X_s$ is algebraic$\}$ is a countable union of 19-dimensional manifolds.

THEOREM 20 (Kodaira)

(1) Any small deformation of rational surface is rational.

(2) Let $M_o$ be an algebraic surface with $C_1^2(M_o) > 0$.
Then any small deformation of $M_o$ is an algebraic surface.

(3) Let $M_o$ be algebraic surface with no exceptional curve of the first kind. If $C_1^2(M_o) \neq 0$, $M_o$ can be deformed into a nonalgebraic surface by small deformations.

(4) Let $M_o$ be birationally equivalent to a ruled surface, then any small deformation $M_s$ of $M_o$ is also an algebraic surface which is birationally equivalent to a ruled surface.

(5) Every compact Kahler surface is a deformation of an algebraic surface.

(6) Every surface with an even Betti number is a deformation of an algebraic surface.

REMARKS

For the notion of ruled surfaces, rational surfaces and exceptional curves of the first kind, see Section 3, Part 2 of Chapter III.

In the above $C_1(M_o)$ represents the first Chern class of $M_o$, and $C_1^2(M_o)$ represents the value of $C_1^2$ on the 4-cycle $M_o$. It can be proved that if $\tilde{C}_1^2(M_o) > 0$ for an analytic surface, then $M_o$ must be algebraic.

(C) Hyperbolicity is locally preserved:

Around 1968, Kobayashi introduced the notion of a hyperbolic complex manifold. This has turned out to be a powerful concept and has led to many important advancements. We now give the definition, cite simple examples and state some problems relevant to our context.

DEFINITION

Let $D = \{z \in C \mid |z| < 1\}$; let $\rho$ be the Poincare-Bergman metric of D; let M be any complex manifold; let p,q be any two points of M. Choose points $p = p_o$, $p_1, \ldots, p_k = q$ in M and points $a_k, \ldots, b_1, \ldots, b_k$, in D and holomorphic maps $f_1, \ldots, f_k$ of D into M such that $f_i(a_i) = p_{i-1}$ and $f_i(b_i) = p_i, i = 1, \ldots, k$. Define

$$d_M(p,q) = \inf \sum_{i=1}^{k} \rho(a_i, b_i)$$

where inf is taken over all possible choices of points and maps. Then $d_M : M \times M \to R$ is a pseudodistance; $d_M$ is a generalization of the Poincare-Bergman metric $\rho$ of D.

We say M is hyperbolic if $d_M$ is a distance function. Every compact Riemann surface is hyperbolic; every bounded domain in $C^n$ is hyperbolic and $C - \{a, b\}$, where a, b are two distinct points of C, is also hyperbolic. But $C^n$ is not hyperbolic as $d_M$ is trivial for $M = C^n$, i.e. $d_{C^n}(Z_1, Z_2) = 0$ for any two $Z_1, Z_2 \in C^n$. It can be proved that for a compact hyperbolic manifold M, the automorphism group Aut(M) is finite.

Kobayashi [472] raised the following problems:

Problem 1: If in a holomorphic family $(X, S, \pi)$, a fibre $M_s = \pi^{-1}(s)$ is hyperbolic, does this imply that $M_t$ is hyperbolic for t sufficiently near s?

Problem 2: If in a holomorphic family $(X, S, \pi)$ all the fibres except one $M_o$ are hyperbolic, does it imply that $M_o$ is also hyperbolic?

The first problem has been solved recently in the affirmative by Wright [980]. Brody and Green [122] have given an example answering Problem 2. It seems to be unknown whether <u>any</u> deformation of a hyperbolic manifold is hyperbolic.

(D) Homogenity is not locally preserved:

Consider a homogeneous compact complex manifold M. Homogeneity of M means that the holomorphic automorphism group of M acts transitively on M. If $\underline{M}$ is a simply connected compact differentiable manifold with nonvanishing Euler characteristic, a fundamental theorem of Wang [951] says that there are only finitely many inequivalent (isomorphism of classes of) homogeneous complex structures on $\underline{M}$; if the Euler characteristic of $\underline{M}$ vanishes, then either there is no homogeneous complex structure on $\underline{M}$ or there are uncountably many inequivalent homogeneous complex structures on $\underline{M}$.

Let us call a simply connected homogeneous complex manifold a Wang manifold (Wang called such manifolds C-spaces). Wang showed that the underlying manifolds of most of the Wang manifolds admits infinitely many non-Kahlerian complex structures. Griffiths[309] studied in great detail deformations of compact homogeneous complex manifolds. Such a manifold M admits two kinds of representations as coset spaces of Lie groups: $M = G/U$ where $G, U$ are complex Lie groups and $M = H/V$ where $H, V$ are compact Lie groups with M semi-simple. Using representation theory, Griffiths constructed a family of deformations for a given compact homogeneous complex manifold and showed that homogeneity need not be preserved even locally. He also observed the occurrence of the jumping phenomenon in this situation.

It is appropriate to mention here the result of Samelson [809] which states that an even dimensional compact semi-simple Lie group always admits a homogeneous complex structure.

## SECTION 5

## DEFORMATION THEORIES FOR OTHER STRUCTURES

Ever since the fundamental work of Kodaira-Spencer, and Kuranishi, their techniques and concepts have been used successfully in diverse situations. In particular we have deformation theories for other structures also. The various deformation theories can be listed as follows:

1) Deformation theory for holomorphic vector bundles over a compact complex manifold.

2) Deformation theory for compact complex spaces and for singularities of complex spaces.

3) Deformation theory for manifolds with boundary.

4) Deformation theory for holomorphic maps.

5) Deformation theory for varieties, schemes and algebraic singularities. (Algebro-geometric deformation theory).

6) Deformation theories for G-structures, Γ-structures and foliate structures. (Differential geometric deformation theory).

7) Deformation theories for Riemannian metrics, and more generally connections (may be regarded as a particular case of 6)

8) Deformation theories for Lie groups, Lie algebras, and of abstract rings, algebras and sheaves. (Algebraic deformation theories). Deformation theory of compact complex manifolds and the theories 1 to 4 could be called analytic geometric deformation theories.

Among these the first one is completely analogous to the Kodaira-Spencer-Kuranishi theory and follows essentially from their work. We give an account of this in this section. We give only the important references for the other theories.

1) Deformation theory of holomorphic vector bundles.

(1) Let X be a compact complex manifold and G be a connected complex Lie group. Let O be a holomorphic principal bundle over X with structure group G. We denote this bundle by $G \to P \to X$. Without changing G, we may deform this bundle by varying the bundle structure P and keeping the complex structure of X fixed, or we may vary both P and X. Here, for simplicity, we consider only holomorphic principal bundles over a fixed compact complex manifold with a fixed structure group G. By a holomorphic family of principal bundle over X parametrized by a connected complex space M, we mean a holomorphic principal bundle P over $X \times M$. For $t \in M$, let $P_t$ be the restriction of P to $X \times \{t\}$. Thus we have a holomorphic family of principal bundles $\{G \to P_t \to X\}_{t \in M}$, which we denote by $P = \{P_t\}_{t \in M}$. Given a bundle $G \to P \to X$, if there exists a family $P = \{P_t\}_{t \in M}$, such that $G \to P \to X$ is isomorphic with $G \to P_{t_o} \to X$ for some $t_o \in M$, we say that the family is a holomorphic family of deformations of the given bundle. A holomorphic family $P = \{P_t\}_{t \in M}$ is said to be complete at $t_o$ M if for any holomorphic family $R = \{R_s\}_{s \in N}$ of deformations of $P_{t_o} \approx R_{s_o}$, $s_o \in N$, there exist an open neighbourhood U of $s_o$ in N and a holomorphic map $f: U \to M$ such that P induces R over $X \times U$ by the map $X \times U \to XM$ given by $(x,r) \to (x,f(r))$.

It is known that for any holomorphic bundle $G \to P \to X$, X compact, there exists a locally complete family of deformations. We now show that the construction of locally complete family of deformations of a principal bundle over a compact complex manifold is analogous to the Kuranishi construction of the locally complete family of deformations of a compact complex manifold.

2) Almost complex principal bundles

Associated with the given bundle, we have the Atiyah sequence

$$O \to L(P) \to Q \to T \to O.$$

Here T is the holomorphic tangent bundle of X, Q the bundle of tangent vectors to P defined along one of its fibres and invariant under G, and $L(P) = P \times g$ the vector bundle with fibre the Lie algebra g of G, associated with P by the adjoint representation of G. Let the exact sheaf sequence corresponding to the previous sequence be

$$O \to \underline{L(P)} \to \underline{Q} \to \underline{T} \to O$$

Let $A^{0,q}$ be the sheaf of germs of scalar-valued $(0,q)$ forms on X and $\Theta$ be the sheaf of germs of holomorphic functions on X. Let

$$A^q = \underline{L(P)} \otimes_{\Theta} A^{0,q}, \qquad B^q = \underline{Q} \otimes_{\Theta} A^{0,q}, \qquad C^q = \underline{T} \otimes_{\Theta} A^{0,q}$$

$$A = \sum_{q \geq 0} A^q, \qquad B = \sum_{q \geq 0} B^q, \qquad C = \sum_{q \geq 0} C^q$$

The exterior differentiation operator $\bar{\partial}$ and the bracket operator $[\ ,\ ]$ can be extended to act on A, B, C and we denote the extended operators also by the same symbols $\bar{\partial}$ and $[\ ,\ ]$. Then, corresponding to the sheaf sequence (    ) we get an exact sequence of Lie algebra complex

$$O \to A \to B \xrightarrow{\pi} C \to O.$$

Let $\underline{X}$ be the underlying differentiable manifold of X and $C\,T\underline{X}$ be the complexified tangent bundle of $\underline{X}$. Recall that an almost complex structure can be defined as follows:

DEFINITION

An almost complex structure on $\underline{X}$ is defined to be a $C^{\infty}$-vector subbundle over C, say T, of $C\,T\underline{X}$ such that we have a direct sum decomposition $C\,TX = T \oplus \overline{T}''$.

Any complex structure on $\underline{X}$ induces canonically an almost complex structure on X. Let $J_o'$ be the underlying almost complex structure tensor of X. An almost complex structure is said to be integrable if it is induced by a complex structure. Thus $J_o'$ is integrable. Let $J_o$ be the underlying almost complex structure of P.

DEFINITION

An almost complex fibre bundle structure $G \to P \to X$ is defined to be a pair $(J, J)$, where J (respectively, J) is an almost complex structure on P (respectively on X) such that

(i) J is G-invariant

(ii) the almost complex structure on P/G induced by J is $J'$ and

(iii) J restricted to a fibre gives the integrable almost complex structure on $\underline{G}$ .

As above the notion of integrable almost complex fibre bundle structures are defined.

DEFINITION

An almost complex principal bundle structure on $G \to P \to X$ is an almost complex structure J on $\underline{P}$ such that

(i) J is G-invariant,

(ii) the almost complex structure on P/G induced by J is $J_o'$ and

(iii) J restricted to a fibre gives the integrable almost complex structure on $\underline{G}$.

In deformation theory, the spaces $A^1, B^1, C^1$ play important roles in view of the following propositions [311].

Proposition 1 There is a one-to-one correspondence between almost complex structures $J'$ on X which are sufficiently close to $J_o'$ and elements $\varphi \in C^1$ which are near zero. The integrability condition is

$$\bar{\partial}\varphi - \frac{1}{2}[\varphi,\varphi] = 0$$

Proposition 2 There is a one-to-one correspondence between almost complex fibre bundle structures $(J, J')$ on $G \to P \to X$, which are sufficiently close to $(J_o, J_o')$ and elements $\Psi \in B^1$ which are close to zero and which satisfy $\pi(\Psi) = \varphi$, where $\varphi$ corresponds to $J'$ . The integrability condition is

$$\bar{\partial}\Psi - \frac{1}{2}[\Psi,\Psi] = 0$$

Proposition 3 There is a one-to-one correspondence between almost complex principal bundle structures J on $G \to P \to X$, which are sufficiently close to $J_o$ and elements $\Theta \in A^1$ which are near zero. The integrability condition is

$$\bar{\partial}\Theta - \frac{1}{2}[\Theta,\Theta] = 0$$

3) Kuranishi Space

Fix a Hermitian scalar producet $(\Theta, \varphi)$ for $\Theta, \varphi \in A^p$ and $\delta$ be the formal adjoint of $\bar{\partial}$. Let $\square = \bar{\partial}\delta + \delta\bar{\partial}$. A form $\Theta$ is said to be harmonic if $\square\Theta = 0$ or equivalently if $\bar{\partial}\Theta = \delta\Theta = 0$. Let $H^p$ be the space of harmonic forms in $A^p$. It is well known that $H^p$ is finite-dimensional. Let $H^{o,p}(A)$ be the p-dimensional cohomology group of complex A with respect to $\bar{\partial}$ and $H^p(X, L(P))$ be the p-dimensional cohomology group of X with coefficients in $L(P)$. Then it is well known that $H^p \approx H^{o,p}(A) \approx H^p(X, L(P))$. Also there exist the harmonic projection operator P and the corresponding Green's operator G, yielding the Hodge-Kodaira decomposition: for any $\Theta \in A^p$,

$$\Theta = P\Theta + \bar{\partial}\delta G\Theta + \delta G\bar{\partial}\Theta$$

Two types of norms that are usually introduced in these spaces are the Holder and the Sobolov norms. Let $\alpha_r$ be an orthonormal base of $H^1(X, L(P))$. Let $\Theta_1(t) = \sum_{r=1}^{m} \alpha_r t_r$, where m is the dimension of $H^1(X, L(P))$, be a fixed element of $H^1(X, L(P))$. Then it is easy to prove the following lemma.

Lemma 1

(i) There is a unique convergent (in any of the two norms) power series solution $\Theta(t)$ of the differential equation

$$\Theta(t) = \Theta_1(t) + \frac{1}{2}\delta G[\Theta(t), \Theta(t)] \qquad (*)$$

(ii) $\Theta(t)$ satisfies the integrability condition if and only if

$$P[\Theta(t), \Theta(t)] = 0$$

We now follow up local completeness theorems which are analogous of the corresponding theorems of Kuranishi given above.

THEOREM 21

Let $G \to P \to X$ be a holomorphic principal bundle over a compact complex manifold X. Let $\alpha_r$ be a base for $H^1(X, L(P))$.

Let $W = \{\Theta_1(t) = \sum_{r=1}^{m} \alpha_r t_r, \; |t| < \epsilon\}$, $K = \{t \in W \mid P[\Theta(t), \Theta(t)] = 0\}$.

Then:

(i) There is a one-to-one correspondence between $\Theta_1 \in W$ and solutions $\Theta$ of equation (*) in Lemma 1 for sufficiently small $\epsilon$.

(ii) The family $\{\Theta(t)\}_{t \in K}$ represents a holomorphic family $P = \{P_t\}_{t \in K}$ of deformations of the given bundle.

(iii) The family $P = \{ P_t \}_{t \in K}$ is locally complete at the reference point.

This theorem can be reformulated in the following form.

THEOREM 21'

There exist a neighbourhood W of the origin in $H^1(X, L(P))$ and a complex analytic map $\Phi : W \to A^1$ such that if $K = \{ t \in W \mid P[\Phi(t), \Phi(t) = 0$, then $\{\Phi(t)\}_{t \in K}$ represents a locally complex family of deformations of the given bundle.

REMARK

For the study of deformations of holomorphic fibre bundles, it suffices to consider the deformations of the associated principal bundles. This is because an elementary result of bundle theory says that two fibre bundles over the same manifold having the same fibre and same group are equivalent if and only if their associated principal bundles are equivalent.

2) The theory of deformations of compact complex spaces has been developed in the last ten years. See Kerner [453 ], [454 ], Schuster [820 ], [821 ]. The central question of existence of Kuranishi space in this case has been proved by Douady [212 ], Forster-Knorr [248 ], Grauert [297 ]. See also Donin [202 ], Palamodov [719 ], [720 ], [721 ], Wavrik [953 ]. For equivariant deformation of complex spaces see Cathelineau [148 ]. Deformations of isolated singularities are discussed in Akahori [16 ], Grauert [296 ], Kuranishi [516 ], Tjurin [916 ]. For more information on singularities of complex spaces see Brieskorn [117 ], [118 ], [119 ], Hirzebruch-Mayer [387 ], Holm [397 ], Kas-Schlessinger [443 ], Laufer [527 ].

3) Deformation theory for manifolds with boundary has been developed systematically by Hamilton [347 ]. See also Stanton [869 ]. Deformations of certain types of non-compact manifolds have been dealt with by Andreotti-Vesentini [31 ], Kiremidjian [461 ], [462 ], Markov-Rossi [569 ].

4) Deformation theory for holomorphic maps has been extensively developed by Horikawa [402 ]. In recent years there has been a lot of work on smooth maps between differentiable manifolds and their singularities; special mention must be made of the work of Arnold [32 ], Mather [576 ], Thom [901 ]. Thom's notion of 'unfolding' can be regarded as a deformation concept and the work of Arnold, Mather and Thom could

be considered to constitute a general deformation theory for smooth maps and their singularities.

5) Following Kodaira-Spencer, Grothendieck developed deformation theory for varieties and schemes in his fundamental papers [329]. Deformation theoretic methods have been successfully applied since then in algebraic geometry. Refer to Artin [38], [40], [41] Laudal [526], Lichtenbaum-Schlessinger [539], Mumford [635], Matsumara-Oort [579], Oort [712], [713], [714], [715], Schlessinger [814], [815], Seshadri [841]. There has been extensive work on algebraic singularities and their deformations; for a nice treatment of deformation theory of singularities see Artin [42] and the references given there.

6) For deformation theory of G-structures and Γ-structures the principal references are Griffiths [310], Grauert [298], Guillemin-Sternberg [335], Moolgavkar [619], Que [956].

7) Deformation theory of connections has acquired significance in view of the recent work of Atiyah and others, on Yang-Mills fields. For general theory of connections see e.g. Kobayashi-Nomizu Vol.I [475]. For deformations of connections and metrics see e.g. Raghunathan [759], Calabi [133]. For recent applications to Physics see e.g. Atiyah-Hitchin-Singer [62], [63], Atiyah-Bott [57].

8) Motivated by the deformation theory of compact complex manifolds, Gerstenhaber [269], Nijenhuis-Richardson [697] have developed deformation theories for abstract rings and algebras, Lie groups and Lie algebras. For some of the applications of deformation theory of Lie algebras to Physics see Pommerrat [741]. Recently Trautman [925], [926] has developed a theory for deformations of Sheaves.

# 3 Classification theory of compact complex manifolds

## PART : 1 CLASSIFICATION OF COMPACT RIEMANN SURFACES

The classification of compact connected Riemann surfaces in terms of the genus g is well known. The genus g of a compact Riemann surface R is given by $g = \dim H^1(R, \Theta_R) = \dim H^0(R, \Theta_R)$ where $\Theta_r$ is the structure sheaf of R and $\Omega_R$ is the sheaf of germs of holomorphic differentials on R. The genus is a birational invariant. Topologically R is a sphere with g handles and g = 1/2 (Betti number of R).

A compact Riemann surface R can be regarded as a non singular algebraic curve over C via the non constant global meromorphic functions on R (which exist by the Riemann existence theorem). Conversely any non singular algebraic curve over C defined by $F(X,Y) = 0$ can be regarded as a compact Riemann surface by regarding Y as an algebraic function of X. Thus the notions of compact surfaces and non singular (smooth) algebraic curves are equivalent. This is one of the pleasant aspects in dimension 1. See Griffiths-Harris [319], Chapter 2. If two compact Riemann surfaces are bimeromorphic, they are biholomorphic.

Let R be a compact Riemann surface of genus g. If g = 0 R is rational (ie. it is $P^1(C)$); if g = 1, R is said to be an elliptic Riemann surface. If $g \geq 2$, R is said to be hyperelliptic if there exists a finite morphism $f: R \to P^2(C)$ of degree 2 (ie. there exists a meromorphic function f on R with exactly two poles). If g = 2, R is hyperelliptic and if $g \geq 3$ there are nonhyperelliptic Riemann surfaces; in fact the 'general' Riemann surface is nonhyperelliptic.

Recall the definition of the canonical class (line bundle) of an algebraic variety. Let $K_R$ be the canonical class of R and $|K_R|$ be the associated linear system. If R is rational, it is easy to see $|K_R| = \emptyset$; if R is elliptic $|K_R| = 0$ and $K_R \neq 0, \emptyset$ if the genus > 1. Hence when g > 1, $|K_R|$ defines the canonical map;
$\Phi_K: R \to \mathbb{P}^n(C)$. We have the following basic fact: (see e.g. Griffiths-Harris [319]).

THEOREM

Let R be a compact Riemann surface of genus g = 1. Denote $K_R$ by K.

Then,

(a) $\Phi_{mK}$ is birational for every $m \geq 3$ and for all g.

(b) $\Phi_{2K}$ is birational for egery $g > 3$.

(c) $\Phi_K$ itself is birational if R is not hyperelliptic.

One of the main aims of classification theory in higher dimensions is to get an analogous theorem on the pluricanonical map $\Phi_{mK}$. Using the linear system $K_R$, we can define another important invariant; the Kodaira dimension of R, denoted by $\kappa(R)$, is defined by

$$\kappa(R) = \begin{cases} -\infty \text{ if } |mK| = \emptyset \text{ for all } m \\ \max_m \dim \emptyset\, mK^{(R)}, \text{ otherwise.} \end{cases}$$

It is easy to check that $\kappa(R)$ is a birational invariant; there are only three possible values for $\kappa(R)$: viz $-\infty$, $0$, $1$. Motivated by the classification theory in higher dimensions, we define R to be of elliptic type if $\kappa(R) = -\infty$, to be of parabolic type if $\kappa(R) = 0$ and to be of hyperbolic ( or general) type if $\kappa(R) = 1$. As per this classification, a rational Riemann surface is of elliptic type, an elliptic Riemann surface is of parabolic type and a Riemann surface of genus $g \geq 2$ is of hyperbolic (or general) type. In higher dimensions, rough classification is sought in terms of the Kodaira dimension.

## PART : 2
## CLASSIFICATION OF COMPACT CONNECTED COMPLEX SURFACES

The classical theory of classification of algebraic surfaces is due to the Italian geometers, see [200]. For a detailed account of this classification see Saferavich [802]. Beginning in 1960, in a series of papers [480][481], Kodaira has extended this theory to complex surfaces (throughout, by a complex surface we mean a compact connected complex manifold of dimension two.) This is a vast field and we content ourselves with stating the main results of Kodaira. For details and proofs, the reader is referred to Kodaira [480][481], Bombieri [96], Saferavich [802]. In the four papers of Kodaira [481], there are 57 theorems and we refer to these theorems by their corresponding numbers, given in these papers.

We remarked already that if two compact Riemann surfaces (curves) are bimeromorphic, then they are complex analytically isomorphic. But this is not true in higher dimensions: two complex surfaces may be bimeromorphically isomorphic without being complex analytically isomorphic. The theory of relatively minimal models ( see Zariski [995]) is motivated to overcome this difficulty. According to this theory any complex surface is obtained from a relatively minimal model by a finite number of blowing ups. Except for rational and ruled surfaces two relatively minimal models are bimeromorphic to each other if and only if they are complex analytically isomorphic. Hence, it is sufficient to classify relatively minimal models.

## SECTION 1

## BIMEROMORPHIC INVARIANTS

Complex surfaces are classified in terms of certain fundamental numerical invariants associated with them. We now define these and state the relations connecting them. By a surface we will mean throughout a compact connected complex manifold of complex dimension 2.

Let S be a surface. $b_r(S)$, $r = 1,2,3,4$, will denote its r-th Betti numbers and $c_R(S)$, $r = 1,2$ its r-th Chern class. We represent any cohomology class $C \in H^4(S, Z)$ by the value $c(S)$ of c on the fundamental cycle of the surface oriented in the natural way with respect to the complex structure of S. $C_1^2(S)$ are thus considered as integers. Let $\Theta_S$ be the structure sheaf of S and $\Omega^r(S)$ be the sheaf of germs of holomorphic r-forms. Let $q(S) = \dim H^1(S,\Theta_S)$, $p_g(S) = \dim H^2(S,\Theta_S)$, $h^{r,s}(S) = \dim H^s(S, \Omega^r(S))$. $q(S)$ is called the irregularity of S, $p_g(S)$ is called the geometric genus of S and $h^{r,s}(S)$ are called the Hodge members of S.

The arithmetic genus $p_a(S)$ of S is the Euler-Poincare characteristic $\chi(S,\Theta_S)$ of S with coefficients $\Theta_S$ : $p_a(S) = \chi(S,\Theta_S) = \sum_i (-1)^i H^i(S,\Theta_S)$. We denote $\chi(S,\Theta_S)$ by $\chi(S)$. Sometimes $\chi(S)-1$ is also called the arithmetic genus of S. Each of the three different concepts of the geometric genus, the irregularity and the arithmetic genus can be regarded as a generalisation of the genus of a compact Riemann surface.

A divisor on S is a finite sum $\Sigma n_i E_i$, $n_i \in Z$ where $E_i$ is an irreducible curve on S. The divisor is called non-negative if all the $n_i$ are non-negative and it is called positive if it is non-negative and not zero.

There exists symmetric non-degenerate pairing $H_2(M,Z) \times H_2(M,Z) \to Z$ since S is an oriented 4 dimensional real manifold. The intersection number of two divisors $D_1, D_2$ is defined to be the intersection number of their fundamental homology class $(D_1)$ and $(D_2)$, and it is denoted by $D_1 . D_2$. If $L_1 \to S$, $L_2 \to S$ are two line bundles, then their intersection number $L_1 . L_2$ is defined to be $[C_1(L_1) \cup C_1(L_2)][S]$. $D_1 . L_1$ is defined to be the value of $C_1(L_1)$ on the fundamental class $(D_1)$. The projective space of non-negative divisors linearly equivalent to D is denoted by $|D|$.

Let $K_S$ denote the canonical line bundle of S. That is $K_S : \overset{2}{\wedge} T^*(S)$ where $T^*(S)$ denotes the cotangent bundle of S. $K_S$ can also be regarded as a divisor in which case it is called the canonical divisor. The first Chern class $C_1(S)$ of S is $-C_1(K)$. The virtual genus $\pi(D)$ of a divisor D on S is defined by

$$2\pi(D) - 2 = D^2 + D.K.$$

If D is effective, $\pi(D)$ is a non-negative integer. If D is non-singular curve on S, then $\pi(D)$ is the usual genus of D.

The following are very important theorems:

THEOREM 1 (Riemann-Roch theorem for divisors (line bundles) on a surface)

For a divisor D on an algebraic surface S,

$$\chi(S,[D]) := \sum_{i=0}^{2} (-1)^i H^i(S,[D]) = 1/2\, D(D-K_S) + 1/12\,(C_1^2(S) + C_2(S)).$$

THEOREM 2 (Hirzebruch Riemann-Roch theorem)

$$\chi(S) = \chi(S, \Theta_S) = 1 - q(S) + p_g(S).$$

THEOREM 3 (Noether's formula)

$$\chi(S) = 1/12\{C_1^2(S) + C_2(S)\}.$$

THEOREM 4 (Serre duality theorem)

$$h^{0,2}(S) = h^{2,0}(S).$$

THEOREM 5 (Hirzebruch signature (index) theorem)

Let $b^+$ and $b^-$ be respectively equal to the number of positive and negative eigenvalues (with due count of multiplicities) of the intersection matrix on $H^2(S,R)$. Then

$$b^+ - b^- = 1/3\,(C_1^2 - 2C_2).$$

From the Hirzebruch signature theorem and Noether's formula we have:

THEOREM 6 (Kodaira's theorem 3)

If $b_1$ is even, then $q = h^{1,0} = \frac{1}{2}b_1$ and $b^+ = 2p_g + 1$. If $b_1$ is odd, $q = h^{1,0} + 1 = \frac{1}{2}(b_1 + 1)$ and $b^+ = 2p_g$.

Let $P_m(S) = \dim|mK_S| + 1$. $P_m(S)$ is called the m-th plurigenus of S. $P_m(S) = \dim H^0(S, mK_S)$. Note $P_1(S) = p_g(S)$ by duality.

THEOREM 7 (Kodaira's theorem 57)

If $P_{12}(S) = 0$, then $P_m(S) = 0$ for all m.

THEOREM 8 (Kodaira)

If $P_m(S) = 0$ and n divides m, then $P_n(S) = 0$.

Let $N(S,K_S) := \{ \text{integers } m \geq 1 \mid P_m(S) \geq 1 \}$. Assume $N(S,K_S) \neq \emptyset$. Let $m \in N(S,K_S)$. Let $\sigma_0, \ldots, \sigma_N$ be a basis of $H^0(S, mK_S)$. With respect to this choice of basis we have a meromorphic mapping ( see Section 1, Part B, Chapter 1)

$$\Phi_{mK_S} : S \rightarrow P^N(C).$$

$\Phi_{mK_S}$ is called the m-th pluricanonical map of S. $R[S,K_S] = \bigoplus_{m \geq 0} H^0(S, mK_S)$ is called the canonical ring of S. It is important to know when $\Phi_{mK_S}$ is holomorphic and when the canonical ring is finitely generated. It is known that $\Phi_{mK_S}$ is holomorphic when the complex linear system $mK_S$ associated to $mK_S$ has no base points and no fixed components ( see for example Ueno [932], Lemma 4.20.1).

THEOREM 9 ( Mumford [632], Kodaira [480])

The canonical ring is finitely generated when S is a nonsingular algebraic surface.

This is not true in general in higher dimensions. As in the case of a Riemann surface, the Kodaira dimension of S (also called rhw canonical dimension of S) denoted by $\kappa(S)$, is defined as follows

$$\kappa(S) := \begin{cases} \max_m \dim \Phi_{mK_S}(S) \text{ if } N(S,K_S) \neq \emptyset \\ -\infty \text{ if } N(S,K_S) = \emptyset. \end{cases}$$

Note that there are only four possible values for $\kappa(S)$, namely 0,1,2 and $-\infty$.

The algebraic dimension of S, denoted by a(s), is the transcendence degree of the field of meromorphic functions of S ( see Section 3 of Chapter 1). Note $a(S) = 0$, 1 or 2; and $\kappa(S) \leq a(S)$ always.

Associated to a complex surface S there exists a complex torus A(S) and a holomorphic map $\alpha : S \to A(S)$ such that for any holomorphic map $\beta : S \to T$ of S into a complex torus T, there exists a unique Lie group homomorphism $h : A(S) \to T$ and a unique element $t \in T$ such that $\beta(x) = h(\alpha(x)) + t$, $x \in S$. $(A(S) = \alpha)$ is called an Albanese torus of S ( For any compact complex manifold M an Albanese torus A(M) exists, see Blanchard [90]). For any variety an Albanese variety which is an abelian variety exists (see Lang [520]). If $(A_1(S), \alpha_1)$ and $(A_2(S), \alpha_2)$ are two Albanese tori for S, then $A_1(S)$ and $A_2(S)$ are isomorphic as Lie groups and there exists a uniquely determined $a_1 \in A_1(S)$ such that $\alpha_1(x) = \alpha_2(x) + a_1$, $x \in S$. Thus an Albanese torus is uniquely determined up to translations and hence we can say 'the' Albanese torus of S. The Albanese dimension of S is defined to be the dimension of A(S) and is denoted by t(S). From the construction of the Albanese torus, it follows that $t(S) \leq h^{1,0}(S)$ and if S is Kahler, we have equality: $t(S) = h^{1,0}(S) = 1/2\, b_1(S)$.

THEOREM 10

For a complex surface S, the numerical characters $q(S)$, $p_g(S)$, $p_a(S)$, $p_m(S)$, $\kappa(S)$, $a(S)$ and $t(S)$ are all bimeromorphic invariants of the surface.

We write $b_r$, $h^{r,s}$, $p_g$, ... for $b_r(S)$, $h^{r,s}(S)$, $p_g(S)$ ... where there is only one surface S under consideration.

## SECTION 2

## BLOWING UPS AND BLOWING DOWNS

Let M be any complex manifold of dimension n. Let x be any point of M. Blowing up M at x means 'replacing' x by the projective space $P^{n-1}(C)$ resulting in a new complex manifold denoted by $\sigma_x M$. This process (also called $\sigma$ - process, quadratic transformation) is carried out as follows. Let U be a coordinate deighbourhood of x in M with local coordinate $Z = (Z_1, \ldots, Z_n) : U \to C^n$ such that $Z(x) = 0$. Consider the map $f : U - \{x\} \to P^{n-1}(C)$ given by $f(x) = [Z_1(x), \ldots, Z_n(x)]$.

Let $\Gamma$ be the graph of f and let $M_x = \{x\} \times \mathbb{P}^{n-1}(C)$. Let $N = \Gamma \cup M_x$. Then N is a non-singular submanifold of $U \times \mathbb{P}^{n-1}(C)$. In fact let $w_1, \ldots, w_n$ be homogeneous coordinates in $\mathbb{P}^{n-1}(C)$ and let $V_i$ be the open subset of $U \times \mathbb{P}^{n-1}(C)$ given by $w_i \neq 0$. Then $V_i$ covers $U \times \mathbb{P}^{n-1}(C)$. We can assume $w_i = 1$ in $V_i$ and consider $Z_1, \ldots Z_n$, $w_1, \ldots, w_{i-1}, w_{i+1}, \ldots, w_n$ to be the local coordinates in $V_i$. With respect to these coordinates N is given by $Z_j = Z_i w_j$ $(1 \leq j \leq n, j \neq 1)$ so that $w_1, \ldots, w_{i-1}, Z_i, w_{i+1}, \ldots, w_n$ give a system of local coordinates in $V_i$. Using the projection $\Gamma \to U - \{x\}$, we identify $U - \{x\}$ with $\Gamma = N - M_x$ in the disjoint union of $M - \{x\}$ and N to get the new manifold $\sigma_x M$. $\sigma_x M$ does not depend on any of the choices we have made but depends only on M and x. We have a projection map $\pi_x : \sigma_x M \to M$ which maps $M_x$ onto x and maps $\sigma_x M - M_x$ biholomorphically onto $M - \{x\}$. This blowing up process can be iterated. Let $x_1 \in M$. Consider $\sigma_{x_1} M$. Pick a point $x_2$ in $M_{x_1} = \pi_{x_1}^{-1}(x_1)$. Now blow up $\sigma_{x_1} M$ at this point to get $\sigma_{x_1 x_2} M := \sigma_{x_2} \sigma_{x_1} M$. $M_{x_1}$ and $M_{x_2}$ intersect exactly at one point and the projection:

$\pi_{x_1 x_2} := \pi_{x_2} \pi_{x_1} : \sigma_{x_1 x_2} M \to M$ maps $M_{x_1} \cup M_{x_2}$ onto $\{x\}$. We now blow up at

a point $x_3 \in M_{x_1} \cup M_{x_2}$ and this process can be continued. After s number of blow ups we get a complex manifold $\bar{M} = \sigma_{x_1 \ldots x_s} M$ and projection $\bar{\pi} : \bar{M} \to M$ with $\pi^{-1}(x_1) = M_1 \cup M_2 \cup \ldots \cup M_s$. $M_i$ and $M_j$ are either disjoint or intersect exactly at one point. Let us now consider the case of a complex surface S. Let $x \in S$. Let $z_1, z_2$ be the local coordinates in U centred at x. In the blow up surface $\sigma_x S$ these coordinates are replaced by two systems u. v and $\bar{u}$, $\bar{v}$ such that $z_1 = u$, $z_2 = uv$ and $z_1 = \overline{uv}$, $z_2 = \bar{v}$. $M_x$ is given by $u = 0$ and $v = 0$. $M_x$ is the 2-sphere embedded in $\sigma_x M$ and represents a cocycle with self intersection number -1. A non-singular rational curve (isomorphic with $\mathbb{P}^1(C)$) is called an exceptional curve of the first kind if its self intersection number is -1. Hence blowing up a point on the surface means replacing that point by an exceptional curve of the first kind, resulting in a non-singular complex surface. Conversely, an exceptional curve of the first kind can be blown down. That is it can be 'replaced' by a point such that the resulting surface is again a non-singular complex surface. The converse was first proved for algebraic surfaces by Castelnuovo and it is well known as Castelnuovo's criterion for contractibility. For complex surfaces this follows from Grauert [295].

See Kodaira [480] for details. More generally we can ask the following question. Given a complex analytic space X and a closed complex analytic set A in X when can we blow down A ? In recent years a number of interesting results have been proved in this direction. See Grauert [295], Moishezon [611], Markoe-Rossi [569], Siu [858], Knorr-Schneider [469], Nakano [666], Artin [34]. An algebraic surface reamins algebraic after blowing up or down, Kodaira [481].

Recall that a proper surjective holomorphic map $f: S_1 \to S_2$ between surfaces is called a modification if there exists analytic subsets $A_1 \subset S_1$ and $A_2 \subset S_2$ such that $f|_{S_1 - A} : S_1 - A \to S_2 - A$ is biholomorphic.

The following theorems are important.

THEOREM 11 (Hopf)

Let $\pi : S \to W$ be a modification. There exist surfaces $S_i$, $i = 0, \ldots, m$ and surjective holomorphic $\pi_i : S_i \to S_{i+1}$, $i = 0, 1, \ldots, (m-1)$ such that

(a) $S_o = S$, $S_m = W$.

(b) $\pi_i : S_i \to S_{i+1}$ is the blow up of $S_{i+1}$ with centre a point $p_{i+1} \in S_{i+1}$.

(c) $\pi = \pi_{m-1} \circ \pi_{m-2} \circ \ldots \circ \pi_0$.

From the above we get the following:

THEOREM 12 (Elimination of points of indeterminacy of rational mappings)

Let $f: S \to W$ be a meromorphic mapping of surfaces. Then there exists a surface $\hat{S}$, a surjective morphism $\pi : \hat{S} \to S$ and $g: \hat{S} \to W$ such that (a) $\pi : \hat{S} \to S$ is obtained by a finite succession of blow ups and (b) $g = f \circ \pi$.

A complex surface is called a relatively minimal model (or relatively minimal surface) if and only if it does not contain any exceptional curve of the first kind. Every complex surface can be obtained from a relatively minimal model S, by a finite number of blowing ups. A surface S is a minimal surface (model) if every surface in the bimeromorphic class of S can be got from S by a finite number of blowing ups.

We have the following basic fact:

THEOREM 13 (Kodaira's theorem 56, Castelnuovo-Enriques-Zariski-Kodaira)

If the bimeromorphic class of a surface S has no minimal model then S is a ruled surface.

REMARK

There exists no theory of minimal models for complex manifolds of dimension greater than two.

## SECTION 3
## SPECIAL AND GENERAL SURFACES

### (1) Ruled Surfaces

A surface S is said to be ruled if $P_n = 0$ for all n. Ruled surfaces are characterised by $P_{12} = 0$. One can show that a surface S is a ruled surface (of genus of g) if and only if S is birationally equivalent to a product of $\mathbb{P}^1(C)$ and a nonsingular curve of genus g. A ruled surface is rational if and only if its genus is equal to zero. A minimal ruled surface of genus $g \geq 1$ is a $\mathbb{P}^1(C)$ bundle over a non-singular curve of genus g. For details see Nagata [658], Maruyama [571].

### (2) Rational Surfaces

A surface S is rational if it is birationally equivalent to $\mathbb{P}^2(C)$. Hence for any rational surface S, $\chi(S) = 1$ and $P_m = 0$ for all m. A minimal rational surface is either $\mathbb{P}^2(C)$ or a $\mathbb{P}^1(C)$ bundle over $\mathbb{P}^1(C)$, see Nagata [658]. To find a good criterion for rationality is a very important problem in algebraic geometry. Castelnuovo [146] was the first to give a criterion for rationality of algebraic surfaces. See also Safarevich [802], Zariski [994], [997]. We give a list of important results. Let S be any nonsingular algebraic surface.

(1) An algebraic surface is rational if and only if it contains an irreducible rational curve C whose linear system $|C|$ is of dimension at least 1 (Noether's Lemma) (see e.g. Griffiths-Harris [319], p.513).

(2) If $q = P_2 = 0$, then there exists on S a nonsingular rational curve E with $E^2 \geq 0$. (Kodaira's Theorem 48).

(3) If $q = P_2 = 0$, then S is rational (Castelnuovo's Criterion for Rationality).

(4) If $q = P = 0$, then S is either $\mathbb{P}^2(C)$ or a rational ruled surface. (Castelnuovo -Andreotti, Kodaira's Theorem 49).

(5) If $q = 0$ and there exists on S a nonsingular rational curve E with $E^2 \geq 0$, then S is rational (Proposition 1.2 Hirzebruch [390]).

(6) If q = 0 and there exists on S an irreducible curve E with $K.E < 0$ and $E^2 \geq 0$, then S is rational (Proposition 1.4 Hirzebruch [390]).

(7) If q = o and there exist two exceptional curves of the first kind which intersect, then S is rational. (Proposition 1.6 Hirzebruch [390]).

(8) If $S_1$ is rational and if $f: S_1 \to S_2$ is a surjective holomorphic map then $S_2$ is rational (Luroth's theorem) (see e.g. Griffiths-Harris [319], p. 541).

(3) K3 Surfaces

A complex surface S is called a K3 surface if and only if its irregularity q is equal to zero and its canonical bundle K is trivial. Equivalently, S is a K3 surface if and only if q and the first Chern class $c_1$ are equal to zero. It is known that any deformation of a K3 surface is a K3 surface ; Kodaira proved (Kodaira's Theorem 13) the Weil-Andreotti conjecture that any K3 surface is a deformation of a nonsingular quadratic surface in $\mathbb{P}^3(C)$ (see Grauert [294]). Hence any K3 surface is simply connected and there is a unique diffeomorphism type of K3 surface. In [488] Kodaira constructed all complex surfaces which have the same homotopy type as a K3 surface. The problem of determining all complex surfaces which have the same homeomorphism type as a K3 surface remains open. The existence of moduli spaces for algebraic K3 surfaces has been proved by Pjateckii-Sapiro and Safarevich [735] and for Kahler K3 surfaces, see Burns and Rapport [128]. Saint Donat [804] has given models for K3 surfaces. It can be proved that a quartic surface in $\mathbb{P}^3(C)$ can have at most 16 double points. A quartic surface with exactly 16 double points is called a Kummer surface. A minimal nonsingular surface which is birationally equivalent to a Kummer surface is a K3 surface.

(4) Elliptic Surfaces

A complex surface is an elliptic surface if it admits at least one elliptic fibering, i.e., a holomorphic map $\varphi$ of S onto a nonsingular curve $\Delta$ such that all but a finite number of fibres are nonsingular elliptic curves. Kodaira [480] has described in detail all possible types of such fiberings. This has stimulated further studies of 'degeneration' ; for some aspects of degeneration see Namikawa [671], Namikawa-Ueno [673], Mumford [639]. If for a surface S, $a(S) = 1$ or $K(S) = 1$, then S has a unique structure of an elliptic surface. The image of the points of S at which $\varphi$ is not of maximal rank is a finite subset $(a_1, \ldots, a_r)$ of $\Delta$. Let $z_i$ be a local coordinate in $\Delta$ around $a_i$ such that $z_i(a_i) = 0$. The divisor D defined by $z_i \cdot \varphi = 0$ is called a singular fibre. The study of the structure of the singular fibres is important in all the degeneration theories. For a general theory of moduli of elliptic fibrations see Deligne-Rapport [188].

(5) Enriques Surfaces

A surface S is an Enriques surface if and only if $q = 0$, $P_2 = 1$ and $p_g = 0$. It is known that an Enriques surface is elliptic and algebraic and its universal covering surface is a K3 surface. For details see Enriques [235], Safarevich [802]. Recently it has been proved that any small deformation of an Enriques surface is Enriques and any two Enriques surfaces are deformations of each other, Horikawa [408].

(6) Hyperelliptic Surfaces

A surface S is a hyperelliptic surface if $q = 1$ and $12\,K$ is trivial (K represents the canonical bundle of S). Any hyperelliptic surface has a finite unramified covering surface which is a product of two elliptic surfaces. The hyperelliptic surfaces can be divided into seven distinct classes. See Griffiths and Harris [319].

(7) Surfaces of Class VII

A surface S is of Class VII if and only if its first Betti number is equal to 1. For a surface of Class VII, its geometric genus $p_g$ is equal to zero (Kodaira's Theorem 26). The class of minimal surfaces with $b_1 = q = 1$ is denoted by $VII_o$. Kodaira has shown that all Hopf surfaces and a certain type of elliptic surfaces belong to $VII_o$. (Kodaira Theorems 27 and 30). Recently Inoue [430], [431] has found new examples of surfaces of Class $VII_o$. It is not known whether there are surfaces of class $VII_o$ other than those given by Kodaira and Inoue. A different presentation of Inoue's example is given in Oda [708].

(8) Surfaces of General Type

A complex surface S is said to be of general type if $\kappa(S) = 2$. A surface of general type S is also characterised by $P_2(S) \geq 1$ and $C_1^2(S) > 0$. It is known that a surface of general type is algebraic, Kodaira [480]. From the classification theorem of Kodaira, it follows that a minimal algebraic surface which is not of general type must be one of the following three classes (1) $\mathbb{P}^2(C)$ and ruled surfaces

(2) Surfaces with trivial canonical bundle (3) Elliptic surfaces, and conversely no surface of any one of these classes can be a surface of general type. It is known that any deformation of a surface of general type is a surface of general type. For studying the structure of surfaces of general type it is necessary to study the finer properties of the map $\Phi_{mK}$. On a minimal surface S of general type there exists only a finite number of rational curves with self-intersection equal to -2. Let E be the set of all such curves and $E_r$, $r = 1,2,\dots,p$, be the connected components of E.

DEFINITION

$\Phi_{mK}: S \to \mathbb{P}^N(C)$ is said to be biholomorphic modulo E if

(i) $\Phi_{mK}$ is everywhere defined on S and is holomorphic

(ii) $\Phi_{mK}$ is biholomorphic on S - E and

(iii) $\Phi^{-1}_{mK}\Phi_{mK}(E_r) = E_r$, for $r = 1,\dots,p$.

We have the following improtant theorems of Bombieri [95], [96].

THEOREM 14 (Bombieri)

Let S be a surface of general type. Then,

(i) $\Phi_{mK}$ is holomorphic for $m \geq 4$ and for $m = 3$ if $C_1^2 > 1$.

(ii) $\Phi_{mK}$ is biholomorphic modulo E for $m \geq 5$ and for $m \geq 4$ if $C_1^2 > 1$ and for $m = 3$ if $p_g = 4$, $C_1^2 > 2$.

Bombieri also obtained information on the nature of possible singularities of $\Phi_{mK}(S)$. From the above theorem it follows that $\Phi_{mK}$ is birational for $m \geq 5$. This is an improvement of the earlier results of Safarevich-Moishezon [802] and Kodaira [487].

Van de Ven [936] proved that for a surface of general type S, the Chern numbers $C_1^2(S)$ and $C_2(S)$ must satisfy the inequality $C_1^2(S) \leq 8C_2(S)$. Recently the inequality has been improved.

(a) $C_1^2(S) \leq 4C_2(S)$ due to Bogomolov.

(b) The best possible result is due to Miyaoka who showed that $C_1^2(S) \leq 3C_2(S)$.

Miyaoka's result has enhanced our knowledge considerably on surfaces of general type.

For global moduli of surfaces of general type see Geiseker [271].

Using the above theorems of Bombieri, Popp [747] has proved that moduli space of surfaces of general type exists as an algebraic space.

The problems of finer classification of surfaces within the class of surfaces of general type has been taken up recently by Horikawa [405]. In a series of papers he has developed a finer classification theory for surfaces of general type. These papers are considered to be very important. See also Van de Ven [937].

## SECTION 4

## CLASSIFICATION THEOREMS

### THEOREM 15 (Classification Theorem 1, Kodaira's Theorem 21)

Relatively minimal surfaces can be classified into the following seven classes:

($I_o$) The class of minimal algebraic surfaces with $p_g = 0$.

($II_o$) The class of K3 surfaces.

($III_o$) The class of complex tori of complex dimension 2.

($IV_o$) The class of minimal elliptic surfaces with $b_1 \equiv 0(2)$. $p_g \geq 1$, $C_1^2 = 0$, $C_1 \neq 0$.

($V_o$) The class of minimal algebraic surfaces with $p_g \geq 1$ and $C_1^2 > 0$.

($VI_o$) The class of minimal elliptic surfaces with $b_1 \equiv 1(2)$. $p_g \geq 1$ and $C_1^2 = 0$.

($VII_o$) The class of minimal surfaces with $b_1 = 1 = q$ and $p_g = 0$.

Table 1

| Class | $b_1$ | $p_g$ | $C_1$ | $C_1^2$ | Structure |
|---|---|---|---|---|---|
| $I_o$ | even | 0 | | | algebraic |
| $II_o$ | 0 | + | 0 | 0 | K3 surface |
| $III_o$ | 4 | + | 0 | 0 | complex tori |
| $IV_o$ | even | + | $\neq 0$ | 0 | elliptic surface of general type |
| $V_o$ | even | + | | + | algebraic surface of general type |
| $VI_o$ | odd | + | | 0 | elliptic |
| $VII_o$ | 1 | 0 | | | certain elliptic surfaces + all Hopf surfaces + Inoue surfaces + ? |

We define classes I, II, ..., VII to be respectively the classes of those surfaces which are birationally equivalent to surfaces belonging to the classes $I_o, II_o, \ldots, VII_o$.

THEOREM 16 (Classification Theorem 2, Kodaira's Theorem 22)

Complex surfaces can be classified into the following I, II, ..., VI characterised by the following conditions:

(I) $b_1 = 0(2)$ and $p_g = 0$.

(II) $b_1 = 0$, $p_g = 1$ and $P_m \sim 1$.

(III) $b_1 = 4$, $p_g = 1$ and $P_m \sim 1$.

(IV) $b_1 = 0(2)$, $p_g > 0$ and $P_m \sim (\frac{1}{2}C_1^2)m$, $C_1^2 > 0$.

(V) $b_1 = 0(2)$, $p_g > 0$ and $P_m \sim (\frac{1}{2}C_1^2)m^2$, $C_1^2 > 0$.

(VI) $b_1 = 1(2)$ and $p_g > 0$.

(VII) $b_1 = 1$.

The classes $II_o$ to $VI_o$ of surfaces are closed under deformations and all the classes I to VII of surfaces are closed under deformations (Kodaira's Theorem 24). Any deformation of any Hopf surface is a Hopf surface and any deformation of an elliptic surface of Class $VII_o$ is either an elliptic surface of Class $VII_o$ or a Hopf surface (Kodaira's Theorems 36 and 27). Note all Hopf surfaces belong to the Class $VII_o$ (Kodaira's Theorem 30).

The following theorem of Kodaira can be considered to be an extension of the classical classification theory of algebraic surfaces.

THEOREM 17 (Classification Theorem 3, Kodaira's Theorem 55)

Relatively minimal surfaces can be classified into the following seven classes.

(1) The class of projective plane and ruled surfaces.

(2) The class of K3 ruled surfaces.

(3) The class of complex tori.

(4) The class of minimal elliptic surfaces with $b_1 = 0(2)$, $P_{12} > 0$, $K \neq 0$.

(5) The class of minimal algebraic surfaces with $P_2 > 0$, $C_1^2 > 0$.

(6) The class of minimal elliptic surfaces with $b_1 = 1(2)$, $P_{12} > 0$.

(7) The class of minimal surfaces with $b_1 = 1$, $P_{12} = 0$.

For the corresponding theorem for algebraic surfaces see the Theorem ROC in Hirzebruch [388].

Table 2

| $b_1$ | $P_{12}$ | $P_2$ | K | $c_1^2$ | Structure |
|---|---|---|---|---|---|
| even | 0 | 0 | | | $\mathbb{P}^2(C)$ or ruled |
| 0 | 1 | 1 | Trivial | 0 | K3 surface |
| 4 | 1 | 1 | Trivial | 0 | complex torus |
| even | + | | Non-trivial | 0 | elliptic surface |
| even | + | + | | + | algebraic surface of general type |
| odd | + | | | 0 | elliptic surface |
| 1 | 0 | 0 | | | certain elliptic surface + Hopf surface + Inoue surface + ? |

Using the above classification theorems Iitaka [423] proved the important fact that all the plurigenera and hence the Kodaira dimension of a complex surface are deformation invariants. Using the above classification theorems we give a classification of surfaces in terms of Kodaira dimension. For a surface S, $\kappa(S) = 2, 1, 0$ or $-\infty$. By definition if $\kappa(S) = 2$, S is of general type and if $\kappa(S) = 1$, S is an elliptic surface of general type. The structure of surfaces with $\kappa(S) = -\infty$ and $\kappa(S) = 0$ are given by:

THEOREM 18 (Classification Theorem in terms of Kodaira dimension)

(a) A minimal surface with $\kappa(S) = -\infty$ is either $\mathbb{P}^2(C)$ or ruled.

(b) A minimal surface with $\kappa(S) = 0$ is either

(i) a K3 surface if $q = 0$, $p_g = 1$,

(ii) an Enriques surface if $q = 0$ and $p_g = 0$,

(iii) an hyperelliptic surface if $q = 1$, or,

(iv) an abelian variety if $q = 2$.

(c) A surface S with $\kappa(S) = 1$ is elliptic.

(d) A surface S with $\kappa(S) = 2$ is (by definition) a surface of general type.

Table 3

| $\kappa$ | $p_g$ | $P_{12}$ | q | $b_1$ | Structure |
|---|---|---|---|---|---|
| $-\infty$ | 0 | 0 | 0 | 0 | rational surface |
| $-\infty$ | 0 | 0 | $\geq 1$ | 2q | ruled surface of genus q |
| $-\infty$ | 0 | 0 | 1 | 1 | surface of class VII |
| 0 | 1 | 1 | 2 | 4 | complex torus |
| 0 | 1 | 1 | 2 | 3 | elliptic surface with K trivial |
| 0 | 0 | 1 | 1 | 2 | hyperelliptic surface |
| 0 | 0 | 1 | 1 | 1 | elliptic surface belonging to class VII with mK trivial , $m > 0$ |
| 0 | 1 | 1 | 0 | 0 | K3 surface |
| 0 | 0 | 1 | 0 | 0 | Enriques surface |

REMARKS

(1) Affine and projective structures on complex surfaces.

We have seen in Section 1, Part B, Chapter 1 that not all compact complex manifolds admit affine or projective structures. The question of which of the complex surfaces admit affine or projective structures and parameterisation of the set of such structures has been investigated especially by Gunning [338], Vitter [941], Ganesh [226].

(2) Classification of complex surfaces by curvature properties has been investigated by many authors. See in particular Blair-Chen [89], Chen [149], Donnelley [205], Hitchin [392], Howard-Smyth [414], Ochiai [707], Yau [985].

## SECTION 5

## CLASSIFICATION OF COMPLETE NONSINGULAR ALGEBRAIC SURFACES IN CHARACTERISTIC $p > 0$

Throughout this subsection X will denote a complete nonsingular (smooth) algebraic surface defined over an algebraically closed field K of characteristic $p > 0$. There exists a theory due to Zariski [995] of relatively minimal modules in this case also. Zariski [994] also proved that Castelnuovo's criterion for rationality holds in this case also. Enriques surfaces in positive characteristic were studied by Artin [33]. Recently Mumford [636], Bombiere-Mumford [99] have extended the classical classification theory of surfaces to any characteristic, thereby unifying the theories of surfaces. For a summary of these results see Van de Ven [938] and Bombieri-Husemoller [98]; we mention here only those important aspects of the Mumford-Bombieri theory where it differs from the classical theory.

We assume that X is minimal i.e., it does not have any exceptional curve of the first kind. Let $\Theta_X$ be the structure sheaf of X, $\Omega^i_X$ the sheaf of i-forms and $K_X$ the canonical class.

(1) In characteristic $p > 0$, the etale cohomology groups $H^i(X, Q_\ell)$, $\ell \neq p$, replace the usual cohomology groups $H^i(X, C)$. Using etale cohomology groups, the Betti numbers $B_i$ and Euler number e are defined as usual and the formula $e = 2 - B_1 + B_2$ continues to hold. The irregularity q of X is defined now to be the dimension of the Albanese variety of X and the formula $2q = B_1$ continues to hold. The important Riemann-Roch theorem, Serre duality and Noether's formula also hold in positive characteristic.

(2) Whereas every holomorphic 1-form on a complex surface is d-closed (Kodaira's Theorem 1, Kodaira [481]), this is no longer true in higher characteristic. In fact the relation $w = df$ may not even hold formally over K. Hence the dimension of $H^o(X, \Omega_X)$ (or $H^o(X, \Theta_X)$) may not be equal to the irregularity q.

(3) Let $\Delta = 2 \dim H^1(X, \Theta_X) - B_1$. For algebraic surfaces in characteristic zero, $\Delta$ is zero. In characteristic p, $\Delta$ is an even positive integer and $0 \leq \Delta \leq 2p_g$ (see Bombieri -Mumford [99]) where $p_g$, as in the classical case, is the geometric genus of X and is given by $p_g = \dim H^{o,2}(X, \Omega_X)$. The numerical invariant $\Delta$ can be taken to measure the 'nonclassical' nature of the theory of classification of surfaces in characteristic p.

(4) Let $h^{p,q}(X) = \dim H^q(X, \Omega^p_X)$. It is true that $h^{o,2}(X) = h^{2,o}(X)$ (Serre Duality) but in general $h^{o,1}(X) \neq h^{1,o}(X)$ (see e.g. Serre [836]).

(5) Another important difference is that there may exist nonintegrable vector fields; that is there may exist vector fields D with no one parameter subgroup of automorphisms such that $Df = \frac{d}{dt} \varphi_t(f) |_{t=0}$. (Refer in this connection to the important paper Rudikov-Shafarevich [800]). Similarly, not every 1-form can be integrated.

(6) The classical Sard's Lemma: "If $f : M \rightarrow N$ is a proper holomorphic map of complex manifolds, there exists an analytic subset $N_1 \subset N$ such that $f^{-1}(x)$ is a

non-singular submanifold of M for $x \notin N_1$" does not hold in general in characteristic $p > 0$. The failure of Sard's Lemma gives rise to pathological situations in characteristics 2 and 3.

(7) The Kodaira vanishing theorem is not true in general in characteristic $p > 0$. (Raynaud [777]).

(8) $B_2(X)$, $B_1(X)$, $C_2(X)$, $\chi(X,\Theta_X)$ are all deformation invariants but $\dim H^1(X,\Theta_X)$, $p_g$ and $\Delta$ are in general only upper semicontinuous under deformations.

(9) As in the classical case, the Kodaira dimension $\kappa(X)$ of X is defined:

$$\kappa(X) = \text{tr.deg}_K \bigoplus_{n=0}^{\infty} H^0(X, \kappa_X^n) - 1$$

$\kappa(X)$ is a birational invariant.

There are four possible values for $\kappa(X)$, $\kappa(X) = -\infty, 0, 1$ or 2. The main classification theorem is:

THEOREM 19 (Mumford [637])

Let X be a minimal projective algebraic surface over an algebraically closed field K. Let $K_X$ be the canonical class of X and $\kappa(X)$ the Kodaira dimension of X. Then

(a) $\kappa(X) = -\infty$ if and only if $K_X C < 0$ for some curve C
(b) $\kappa(X) = 0$ if and only if $12 K_X = 0$
(c) $\kappa(X) = 1$ if and only if $K_X^2 = 0$, $K_X D \geq 0$ for all non-negative divisors D.
(d) $\kappa(X) = 2$ if and only if $K_X^2 > 0$ and $K_X D \geq 0$ for all non-negative divisors D.

(10) The structure of surfaces belonging to each of the above four classes have been investigated by Bombieri-Mumford.

(a) Surfaces for which $\kappa(X) = -\infty$ are characterized as rational surfaces.
(b) Surfaces for which $\kappa(X) = 0$ are divided up to the following four classes.
   (1) $B_1 = 0$, $B_2 = 22$, $\chi(X,\Theta_X) = 2$ (these are called K3-surfaces)
   (2) $B_1 = 0$, $B_2 = 10$, $\chi(X,\Theta_X) = 1$ (these are called Enriques surfaces)
   (3) $B_1 = 4$, $B_2 = 6$, $\chi(X,\Theta_X) = 0$ (these are abelian surfaces)
   (4) $B_1 = 2$, $B_2 = 2$, $\chi(X,\Theta_X) = 0$ (these are called hyperelliptic).

Among (2) the non-classical ones (for which $q = 1$, $\Delta = 2$) have been found to occur only in characteristic 2 and among (4) 'the non-classical ones' (for which $q = 2, \Delta = 2$) have been found to occur only in characteristics 2 and 3 and they are called quasi hyperelliptic.

(c) Surfaces for which $\kappa(X) = 1$ are called elliptic; the non-classical ones, called quasi elliptic have been found only in characteristic 2 and 3.

(d) Surfaces for which $\kappa(X) = 2$ are called surfaces of general type. Their structures and finer subdivisions have been extensively studied.

## SECTION 6

## HIRZEBRUCH'S CLASSIFICATION THEORY OF HILBERT MODULAR SURFACES

Around 1970, Hirzebruch started the programme of classifying Hilbert modular surfaces and has essentially completed the programme recently. This is considered to be a major development of this decade. We give here a very brief summary of Hirzebruch's work. The references for this subsection are Hirzebruch [383], [384], Hirzebruch-Van de Ven [388], Hirzebruch-Zagier [389], [390].

Let $\underline{H}$ be the upper half of the complex plane. Consider the real quadratic field $F = Q(\sqrt{p})$ where p is a prime congruent to 1 mod 4. Let $\Theta$ be its ring of integers and $SL_2(\Theta) = \{\begin{pmatrix} a & b \\ c & d \end{pmatrix} \mid a,b,c,d \in \Theta \text{ and } ad - bc = 1\}$. $SL_2(\Theta)$ has a natural action on $\underline{H}^2$ by

$$\begin{bmatrix} a & b \\ c & d \end{bmatrix} (z_1, z_2) = \left[\frac{az_1+b}{cz_1+d}, \frac{a'z_2+b'}{c'z_2+d}\right]$$

where $a', b', c', d'$ are conjugates of $a, b, c, d$ respectively over Q. The group $SL_2(\Theta)/\{+1,-1\}$ acts effectively on $\underline{H}^2$. This group, denoted by H, is called the Hilbert modular group. The quotent space $\underline{H}^2/H$ is a non-compact surface with singularities. It is known that there are only finitely many singularities and each of them is a quotent singularity. Each singularity is locally the quotent of $C^2$ by the linear action of the group of r-th roots of unity $\xi^r = 1$ for a suitable r. The action is given by $(z_1, z_2) = (\xi z_1, \xi^q z_2)$ where q and r are relatively prime. Prestel [752] has shown that if $a_r$ is the number of singularities of order r in $\underline{H}^2/H$, then (1) for $p = 5$, $a_2 = a_3 = a_5 = 2$, $a_r = 0$ for $r > 5$ and (2) for $p > 5$, $a_2$ = class number of the field $Q(\sqrt{-p})$; $a_3$ = class number of the field $Q(\sqrt{-3p})$ and $a_r = 0$ for $r \geq 5$.

The Euler number $\underline{H}^2/H$ is given in terms of these singularities and the Zeta function of the field F. Precisely we have

$$e(\underline{H}^2/H) = 2\zeta_F(-1) + \sum_{r\geq 2} a_r \left(\frac{r-1}{r}\right).$$

Hirzebruch has shown how to resolve the quotent singularities of $\underline{H}^2/H$ to get a non-compact nonsingular complex surface denoted by $X(p)$. Then he compactifies $X(p)$ by finitely many cusps in a canonical way. The exact number of cusps required is equal to the number of ideal classes of $\Theta$ in F. Let $\overline{X(p)}$ be the compactification of $X(p)$. $\overline{X(p)}$ is now a compact complex surface with finite number of singularities. The number of singularities on $\overline{X(p)}$ is equal to the class number of the field F. Hirzebruch [383] then shows that these singularities admit a 'cyclic' resolution. Resolving these singularities of $\overline{X(p)}$ by this particular method he gets a compact nonsingular complex surface $Y(p)$. $Y(p)$ are called the Hilbert modular surfaces. All $Y(p)$ are regular (their irregularity $q(Y(p)) = 0$) and algebraic. The main theorem of Hirzebruch is the following.

THEOREM 20 (Hirzebruch classification theorem)

Let p be a prime congruent to 1 mod 4, $f = Q(\sqrt{p})$ and $Y(p)$ be the corresponding Hilbert modular surfaces. Then:

(a) $Y(p)$ are rational for $p = 5, 13, 17$

(b) $Y(p)$ are blow ups of elliptic K3-surfaces for $p = 29, 37, 41$

(c) $Y(p)$ are 'honestly' elliptic surfaces (i.e. Kodaira dimension of $Y(p) = 1$) for $p = 53, 61, 73$.

(d) $Y(p)$ are surfaces of general type for all $p \geq 89$.

## PART 3
## CLASSIFICATION OF HIGHER DIMENSIONAL COMPACT COMPLEX MANIFOLDS

In this section we give a brief introduction to the work of Iiataka and Ueno. For details and proofs the reader is referred to Ueno [932]. All complex manifolds that are considered in this Part are assumed to be compact and connected. All algebraic varieties are assumed to be complete, irreducible and defined over complex numbers. By a complex variety we mean a compact irreducible reduced complex space. An algebraic variety can be regarded as a complex variety. (See for example,

Safarevich [802], Chapter VIII and Section 1, Part 1 of Chapter 1 of this volume ).

Kodaira Dimension of a Complex Variety

Let M be a complex manifold of dimension n. As in the case of a surface, the Kodaira dimension of M is defined as follows. Let $K = \bigwedge^n T^* M$ be the canonical line bundle of M. The m-genus of M, denoted by $P_m(M)$, is the dimension of $H^0(M, mK)$ where $mK = K \otimes \ldots \otimes K$ (m times). Set $N(M,K) = \{\text{integers } m \geq 1 \mid P_m(M) \geq 1\}$. Assume $N(M,K) \neq \emptyset$. Let $m \in N(M,K)$. Then with respect to a choice of basis of $H^0(m, \underline{mK})$, we have a meromorphic mapping, called the m-th pluricanonical map, $\Phi_{mK} : M \to \mathbb{P}^N(C)$. The Kodaira dimension of M, denoted by $\kappa(M)$, is defined by

$$\kappa(M) = \begin{cases} \max \dim \Phi_{mK}(M), & \text{if } N(M,K) \neq \emptyset \\ -\infty & \text{if } N(M,K) = \emptyset . \end{cases}$$

For a smooth algebraic variety the m-genus and the Kodaira dimension are similarly defined. The m-genus and the Kodaira dimension of complex manifolds (smooth algebraic varieties) are bimeromorphic (birational) invariants.

Let V be a complex variety (algebraic variety) and V be a nonsingular (smooth) model for V. (Such models exist, Hironaka [365]). Since any two nonsingular (smooth) models of a complex variety (an algebraic variety) are bimeromorphic (birational), we can relate the m-genus and the Kodaira dimension of V by $P_m(V) = P_m(V^*)$ and $\kappa(V) = \kappa(V^*)$.

Note

1) There are (n + 2) possible values for $\kappa(M)$ : $\kappa(M) = -\infty$ and $0 \leq \kappa(M) \leq n$.

2) $\kappa(M) \leq a(M)$ always, where $a(M)$ denotes the algebraic dimension of M. If $a(M) = 0$, then $\kappa(M)$ must be either 0 or $-\infty$. If $\kappa(M) = n$, then $a(M) = n$ and hence M must be Moishezon.

3) Rational and ruled varieties have Kodaira dimension $-\infty$. More generally if the tangential bundle of an algebraic variety is ample, then its Kodaira dimension is $-\infty$.

4) Complex tori have Kodaira dimension zero. More generally if the canonical bundle of a complex manifold is trivial, then its Kodaira dimension is zero.

We mention below two important properties of Kodaira dimension.

Asymptotic behaviour

In the classification theory it is of great importance to know the existence and nature of the limit $P_m(M)/m^{\kappa(M)}$ as m ranges over N(M,K). The following two results give information in this direction. Let M be a complex manifold and d the largest common divisor of the integers in N(M,K). Then,

(i) There exist positive numbers $\alpha$ and $\beta$ and a positive integer $m_o$ such that for any integer $m \geq m_o$, we have

$$\alpha . m^{\kappa(M)} \leq P_{md}(M) \leq \beta . m^{\kappa(M)}$$

(ii) For any positive integer p, there exists a positive number $\delta$ and a positive integer $m_1$ such that for any integer $m \geq m_1$ such that we have

$$P_{md}(M) - P_{(m-p)d}(M) \leq \delta . m^{\kappa(M)-1}$$

Addition Formula

The following important theorem holds:

THEOREM 21

Let $\pi : V \to W$ be a holomorphic fibre bundle over a complex manifold W whose fibre and structure group are respectively a Moishezon manifold M and its automorphism group Aut M. Then $\kappa(V) = \kappa(M) + \kappa(W)$.

This theorem is not true without the assumption on M. In this special case it affirms the following conjecture of Iitaka.

Conjecture $C_{n,m}$

Let f: $V \to W$ be a fibre space where V and W are algebraic manifolds with dim V = n, dim W = m. Then $\kappa(V) \geq \kappa(V_b) + \kappa(W)$, where $V_w = \pi^{-1}(w), w \in W$ is a general fibre of f.

Using the classification theory of surfaces given in the last Chapter it can be checked that conjecture $C_2$ is true for algebraic as well as complex surfaces. Recently Viehweg [940] has proved the conjecture $C_{n,n-1}$.

Classification of Complex Varieties in Terms of Kodaira Dimension

DEFINITION

A complex variety V is called a variety of hyperbolic type (parabolic type, elliptic type respectively) if $\kappa(V) = \dim V$ ($\kappa(V) = 0$, $\kappa(V) = -\infty$ respectively).

If $\kappa(V) > 0$, then according to the fundamental theorem on the pluricanonical fibrations, given below, there exists a projective manifold $V^*$, bimeromorphically equivalent to V, which has the structure of a fibre space whose general fibres are of Kodaira dimension zero. Consider the algebraic variety $W_m = \delta\,\Phi_{mK}(V)$, $m \in N(V,K)$. Let $C(W_m)$ be the function field of $W_m$. Then it can be proved that there exists an integer $m_o$, such that for any integer $m \geq m_o$, $m \in N(V,K)$, $C(W_m) = C(W_{m_o})$ and $C(W_{m_o})$ is algebraically closed in $C(V)$.

<u>THEOREM 22 (Theorem on the Pluricanonical Fibrations, Iitaka [424], Ueno [932])</u>

Let V be an algebraic variety of positive Kodaira dimension. Then there exist a projective (complex) manifold $V^*$, a projective manifold $W^*$ and a surjective morphism $f: V^* \to W^*$, such that

1) $V^*$ is birationally (bimeromorphically) equivalent to V
2) $\dim W^* = \kappa(V)$
3) For a non-empty dense subset U of $W^*$ (in the complex topology), each fibre $V^*_w = \pi^{-1}(w)$, $w \in U$, is irreducible and nonsingular and $\kappa(V^*_w) = 0$ for each $w \in U$.
4) The fibre space $f: V^* \to W^*$ is unique up to birational (bimeromorphic) equivalence. More precisely if $f^{\#}: V^{\#} \to W^{\#}$ is a fibre space satisfying the above conditions, then there are birational (bimeromorphic) maps $g: V^* \to V$ and $h: W^* \to W$ such that $f^{\#} \circ g = h \circ f$, the fibre space $f: V \to W$ is birationally (bimeromorphically) equivalent to a fibre space associated to the pluricanonical map $\Phi_m: V \to \Phi_m(V)$ $(\subset \mathbb{P}^N(C))$ for large m with $P_m > 0$.

One natural question that arises regarding the above theorem is whether we can take the dense set U to be open. From the proof of the theorem it can be seen that if Kodaira dimension is deformation invariant then U can be taken to be open. We already stated the result of Iitaka that Kodaira dimension is deformation invariant for complex surfaces. But contrary to the conjecture of Moishezon [612], the Kodaira dimension for manifolds of dimension $\geq 3$ are not necessarily deformation invariants (see Nakamura [662]). For U to be taken to be open, it is sufficient if the Kodaira dimensions are upper semicontinuous under small deformations. (See Libermann-Serenese [548]).

From the theorem on pluricanonical maps it is clear that the study of classification of complex varieties reduces to (a) the study of complex varieties of hyperbolic, parabolic and elliptic type and (b) the study of fibre spaces whose general fibres are of parabolic type. Associated with the pluricanonical map, we have a fibre space and associated with the Albanese dimension we can introduce a fibre space. Using these fibre spaces, classification of varieties can be studied. For detailed accounts of this programme and for open problems in this area of active current research, the reader is referred to Ueno [932], Popp [749] and the references given there.

The main points are:

1) In the rough classification of parabolic manifolds, the main problem is the settling of the following conjecture of Iitaka-Ueno.

Conjecture $K_n$

(i) If V is of parabolic type, the algebraic map $\alpha : V \to \mathrm{Alb}(V)$ is surjective and has connected fibres.

(ii) The fibre space $\alpha : V \to \mathrm{Alb}(V)$ is birationally equivalent to a fibre bundle over V whose fibre and the structure group are an algebraic manifold F of parabolic type and Aut (F), respectively (equivalence in the etale topology).

By the classification theory of surfaces $K_2$ is true. Ueno [932] has shown that for parabolic manifolds the map $\alpha$ is surjective. He has shown also that $K_n$ is true for all generalised Kummer varieties.

2) In the rough classification of elliptic manifolds, the main problem is the settling of the conjecture $C_{n,m}$. If $C_{n,m}$ is true, then it can be proved that the study of complex manifolds reduces to the study of complex manifolds with $q(M) = 0$ and to the study of fibre spaces where general fibre is of elliptic type.

3) Regarding compact complex manifolds of hyperbolic type, we have the following problems.

Problem 1. Is any deformation of a complex manifold of hyperbolic type a complex manifold of hyperbolic type ? (Recall the theorem of Kodaira that any deformation of a surface of general type is a surface of general type.)

Problem 2. Let M be a complex manifold of hyperbolic type. Does there exist a positive integer $m_o$ depending only on the dimension of M, such that for any integer $m > m_o$, the pluricanonical map $\Phi_{mK}$ is bimeromorphic ? (Recall the corresponding theorem of Bombieri -Kodaira for surfaces of general type.)

Problem 3. Does there exist a moduli space for complex manifolds of hyperbolic type ? If it exists, is it an algebraic space ? (Recall the corresponding theorem of Popp for surfaces of general type.)

Popp [749] has shown that for the rough classification of compact manifolds, the fine (moduli) classification of lower dimensional manifolds is necessary. An excellent reference to the recent developments are the lectures of Popp [749].

# References


1. Abe, K. On a class of Hermitian manifolds, Invent. Math. 51 (1979), 103-120.

2. Abhyankar, S .S. Local Analytic Geometry, Academic Press ( 1964).

3. Abhyankar, S .S. Resolution of Singularities of Arithmetical Surfaces, Arithmetical Algebraic Geometry, Harper and Row (1965), 111-152.

4. Abhyankar, S. S. On the problem of rcsolution of singularities, Proc. Int. Congr. Math. Moscow (1966), 469-481.

5. Abhyankar, S. S. Resolution of Singularities of Embedded Algebraic Surfaces, Academic Press (1966).

6. Abhyankar, S. S. On the birational invariance of arithmetic genus, Rend. Mat. e. Appl. (5) 25 (1966), 77-86.

7. Abhyankar, S. S. Historical ramblings in algebraic geometry, Am. Math. Soc. 83(1976), 409-448.

8. Abikoff, W. Degenerating families of Riemann surfaces. Ann. Math. 105(1971), 29-44.

9. Adams, F. Vector fields on spheres. Ann. Math. 75(1962), 603-632.

10. Adler, A. Integrability of almost complex structures, Mich. Math. J. 13(1966), 499-505.

11. Agostov, M. K. Twenty years of differential topology: Some highlights. Math. Chronicle 4(1976), 77-89.

12. Ahlfors, L. The Complex Analytic Structure of the Space of Closed Riemann Surfaces, Analytic Functions. Princeton Univ. Press (1960), 349-376.

13. Ahlfors, L. Teichmuller Spaces, Proc. Int Congr. Math. (1962), 3-9.

14. Ahlfors, L, et. al. Advances in the Theory of Riemann Surfaces. Ann. Math. Studies 66, Princeton Univ Press (1971).

15. Akahori, T. Intrinsic formula for Kuranishi. Publn. R.I.M.S., Kyoto Univ. 14(1978), 615-641.

16. Akahori, T. Complex analytic construction of the Kuranishi family on a normal pseudo complex manifold. Publn. R.I.M.S. Kyoto Univ. 14(1978), 789-847.

17. Akahori, T., Namba, M. Examples of obstructed holomorphic maps. Proc. Japan Academy 54 Ser.A. No.7(1978), 189-191.

18. Akao, K. Complex structures on $S^{2p+1} \times S^{2q+1}$ with algebraic codimension 1. Complex Analysis and Algebraic Geometry (In honour of Kodaira), Edited by Baily and Shoda, Iwanami Shoten and Camb. Univ. Press, (1977), 2:5-225.

19. Akizuki, Y., Nakano, S. Note on Kodaira-Spencer proof of Lefchetz theorem. Proc. Japan. Acad. 30(1954), 266-272.

20. Alexander, J.W. A proof of invariance of certain constants of analysis situs. Trans. Am. Math. Soc. 16(1915), 148-154.

21. Altman, A. Kleiman, S. Introduction to Grothendieck Duality. Springer Lecture Notes No. 146(1970).

22. Amemiya, I. Lie algebra of vector fields and complex structure. J. Math. Soc. Japan 27(1975), 545-549.

23. Amemiya, I., Masude, K. Shiga, K. Lie algebras of differential operators. Osaka J. Math. 12(1975), 139-172.

24. Andre, M. Methode simpliciale en algebre homologique et algebre commutative. Lecture Notes No. 32 Springer Verlag (1967).

25. Andreotti, A. Recherches sur les surfaces irrugulieres. Acad. R. Belg., Cl. Sci. Mem. Coll. 84 and 87 (1952).

26. Andreotti, A. On the Complex Structure of a Class of Simply Connected Manifolds, Algebraic Geometry and Topology, Princeton Univ. Press (1957).

27. Andreotti, A., Frankel, T. The second Lefchetz theorem on hyperplane sections. Global Analysis (In honour of Kodaira), Princeton Univ. Press and University of Tokyo Press (1961).

28. Andreotti, A., Grauert, H. Theorems de finitude por le cohomologie des espaces complexes. Bull. Soc. Math. France 90 (1962), 193-259.

29. Andreotti, A., Narasimhan, R. Oka's Hoftungslemma and the Levi Problem for complex spaces. Trans Am. Math. Soc. 111(1964), 354-366.

30. Andreotti, A., Stoll, W. Analytic and Algebraic Dependence of Meromorphic Functions. Springer Lecture Notes No. 234 (1971).

31. Andreotti, A., Vessentini, E. On the pseudo rigidity of Stein manifolds. Ann. Scuola Norm. Sup. Pisa 3, 16(1962), 213-223.

32. Arnold, V.I.
(1) Singularities of Smooth Mappings. Russian Math. Surveys 23(1968), 1-43.
(2) Critical points of smooth functions. Proc. Int. Cong. Math. Vancouver. Vol. 1(1974), 19-40.
(3) Critical points of smooth functions and their normal forms. Russian Math. Surveys 305 (1975), 1-75.
(4) Local normal forms of functions. Invent. Math. 35(1976), 87-109.

33. Artin, M. On Enriques Surfaces. Ph.D. Thesis, Harvard University (1960).

34. Artin, M. Some numerical criteria for contractibility of curves of algebraic surfaces. Am. J. Math. 84(1962), 485-496.

35. Artin, M. The etale cohomology of schemes. Proc. Int. Congr. Math. Moscow (1966), 44-56.

36. Artin, M. On the solutions of analytic equations. Invent. Math. 5(1968), 277-281.

37. Artin, M. Algebraic approximation of structures on complete local rings. Pub. Math. IHES 36(1969), 23-58.

38. Artin, M. Algebrization of formal moduli I, Global Analysis, Princeton Univ. Press (1970); II Ann. Math. 91(1970), 88-135.

39. Artin, M. Algebraic Spaces. Yale Univ. Press (1970).

40. Artin, M. Theorems de representabilitie pour les espaces algebriques. Les Presses de Univ. de Montreal (1973).

41. Artin, M. Versal deformations and algebraic stacks. Invent. Math. 27(1974).

42. Artin, M. Deformation of singularities. Lecture Notes, Tata Institute of Fundamental Research, Bombay (1976).

43. Artin, M., Mumford, D. Some elementary examples of unirational varieties which are not rational. Proc. Lond. Math. Soc. Ser. 3. 25(1972), 75-95.

44. Atiyah, M.F. Vector bundles on an elliptic curve. Proc. Lond. Math. Soc. 7(1957), 414-452.

45. Atiyah, M.F. Complex analytic connections in fibre bundles. Trans Am. Math. Soc. 85(1957), 181-207.

46. Atiyah, M.F. Some examples of complex manifolds. Bonner Math. Schriften 6(1958).

47. Atiyah, M.F. Bordism and Cobordism. Proc. Camb. Phil. Soc. 57(1961), 200-208.

48. Atiyah, M.F. Vector fields on spheres. Ann. Math. 75(1962), 603-632.

49. Atiyah, M.F. K-Theory. Benjamin, New York (1964).

50. Atiyah, M.F. Signature of fibre bundles. Global Analysis, Kodaira Volume, Princeton Univ. Press (1969), 73-84.

51. Atiyah, M.F. Eigenvalues and Riemannian Geometry, Manifolds. Tokyo(1970), Tokyo Univ. Press (1975), 5-9.

52. Atiyah, M.F. Riemann surfaces and spin structures. Ann. Scient. Ecole Norn. Sup. 4(1971), 47-62.

53. Atiyah, M.F. Vector Fields on Manifolds. Arbertsgemeinshaft fur Forshung des Landes Nordrheinwestfallen. Heft 200.

54. Atiyah, M.F. Elliptic operators, discrete groups and von Neumann algebras. Asterique 32-33, Societes Math. France (1976), 43-72.

55. Atiyah, M.F. On the geometry of Yang-Mills fields. Lecture Notes in Physics, Springer 80(1978), 216-221.

56. Atiyah, M.F., Bott, R., Patodi, V.K. On the heat equation and index theorem. Invent. Math. 19(1973), 279-330.

57. Atiyah, M.F., Bott, R. Yang -Mills connections ( to appear).

58. Atiyah, M.F., Dupont, J.L. Vector fields with finite singularities. Acta Math. 128(1972), 1-40.

59. Atiyah, M.F., Hirzebruch, F. Riemann Roch Theorem for differentiable manifolds. Bull. Am. Math. Soc. 65(1959), 276-281.

60. Atiyah, M.F., Hirzebruch, F. Vector bundles and homogeneous spaces. Proc. Sym. Pure Math. III (1961), 1-38.

61. Atiyah, M.F., Hitchin, N., Drinfeld and Manin, Yu. Physics Letters 65(A) No.3. (1978), 185-187.

62. Atiyah, M.F., Hitchin, N., Singer, I.M. Deformations of Instantons. Proc. Nat. Acad. Sci. USA 74(1977), 2662 - 2663.

63. Atiyah, M.F., Hitchin, N., Singer, I.M. Self duality in 4 dimensional Riemann Manifolds. Proc. Roy. Soc. Lond. 362(1978), 425.

64. Atiyah, M.F., Patodi, V.K., Singer, I.M. Spectral asymmetry and Riemann geometry. Bull. Lond. Math. Soc. 5(1973), 229-234.

65. Atiyah, M.F., Ward, R. Instantons and Algebraic Geometry. Communications in Math. Phys. 55(1977), 117-124.

66. Aubin, T. Equations du type Monge-Ampere Sur les Varietes Kahlereunes Compactes. C.R. Acad. Sc. Paris 203 (1976), 119-121.

67. Baily, W. L. Jr. On the moduli of Jacobian varieties. Ann. Math. 71(1960), 303-314.

68. Baily, W. L. Jr. On the moduli of Jacobian varieties and curves. Contributions to Function Theory. T.I.F.R. Bombay, 50-62 (1960).

69. Baldassarri, M. Algebraic Varieties. Erg. Math. Springer 1956.

70. Banica, C., Stansila, O. Algebraic Methods in the Global Theory of Complex Spaces. John Wiley, New York (1976).

71. Barratt, M.G., Mahowald, M.E. (Editor) Geometric Applications of Homotopy Theory. Evanston 1977, I and II, Lecture Notes Nos. 657, 658, Springer Verlag.

72. Barth, W. Der Abstand von siner algebraischen Manniglaltikeiten im Komplex-Projective Raum. Math. Ann. 187(1970), 150-162.

73. Barth, W. Moduli of vector bundles on the projective plane. Invent. Math. 42(1977), 63-91.

74. Bass, H. Algebraic K-Theory. Benjamin, New York (1968).

75. Baues, H. J. Obstruction theory of homotopy classification maps. Lecture Notes 628, Springer Verlag (1977).

76. Beauville, A. Surfaces Algebriques Complexes. C.I.M.E. (1977), Asteriques 54 (1978).

77. Becker, J. Parametrisation of analytic varieties. Trans. Am. Math. Soc. 183(1973), 265-292.

78. Berger, M. Les Varieties Riemaniennes a courboure positives. Bull. Soc. Math. France 87(1959), 285-292.

79. Berger, M. On the geometry of the Laplace operator. Differential Geometry, Proc. Symp. Pure Math. AMS, 27(1975).

80. Berger, M., Guduchon, P., Mazet, E. Le spectred'une Riemanienne. Lecture Notes 194, Springer Verlag (1971).

81. Bernard, M. Sur la geometric differentiable de G-structures. Ann. Inst. Fourier (Grenoble) 10(1960), 151-270.

82. Bers, L. Spaces of Riemann surfaces. Proc. Inst. Congr. Math. Edinburgh (1958), 349-361.

83. Bers, L. Lectures on Moduli of Riemann Surfaces. E.T.H. Zurich (1964).

84. Bers, L. Uniformization, Moduli and Kleinian Groups. Bull. London Math. Soc. 4 (1972), 257-300.

85. Bers, L. Spaces of Degenerating Riemann Surfaces. Ann. of Math. Studies 79(1974), 43-55.

86. Bers, L. Quasi conformal mappings with applications to differential equations, function theory and topology. Bull. Am. Math. Soc. 83, 6(1977), 1083-1100.

87. Bishop, E. Mappings of partially analytic spaces. Am. J. Math. 83(1961), 209-242.

88. Bishop, R., Goldberg, S. On the topology of positively curved Kahler manifolds. Tohoku Math. J. 15(1963), 359-364.

89. Blair, D. E., Chen, B. Y. (Editors) Proc. Conf. Differential Geometry. Mich. State Univ. USA (1976).

90. Blanchard, A. Sur les varietes analytique complexes. Ann. Sc. Ecole Norm. Sup., 73(1956), 157-202.

91. Blocks, S., Giesekar, D. The positivity of the Chern classes of an ample vector bundle. Invent. Math. 12(1972 ), 112-117.

92. Blumenthal, O. Uber modulfunktionen von mihroren Veranderlichen. Math. Ann. 56(1903), 509-548.

93. Bochner, S., Yano, K. Curvature and Betti Number. Ann. Math. Studies 32(1957).

94. Bogomolov, R. To appear.

95. Bombieri, E. The pluricanonical map of a complex surface. Int. Collo. Several Complex Variables. Maryland, Lecture Notes 155, Springer Verlag (1970), 35-87.

96. Bombieri, E. Canonical models of surfaces of general type. Publ. Math. IHES, Buses-sur-Yvette 42(1973), 171-219.

97. Bombieri, E., Catanese, F. The tricanonical map of a surface with $k^2=2$, $p_g = 0$. In C. P. Ramanujan: A Tribute, Tata Institute Bombay (1978), 279-290.

98. Bombieri, E., Husemuller, D. Classifications and embeddings of surfaces, Algebraic Geometry, Arcata (1974), Proc. Symp. Pure Math. 29(1975), 327-420.

99. Bombieri, E., Mumford, D. Enriques classification of surfaces in characteristic p, II
Complex Analysis and Algebraic Geometry ( In honour of Kodaira), Iwanami Shoten and Camb. Univ. Press (1977), 23-42, III, Invent. Math. 35(1976), 197-232.

100. Borel, A. Topology of Lie groups and characteristic classes. Bull. Am. Math. Soc. 61(1955), 397-432.

101. Borel, A. Compact Clifford Klein forms of symmetric spaces. Topology 2(1963), 111-122.

102. Borel, A., Haefliger, A. La classe d'homologie fundamental d'une espace analytique. Bull. Soc. Math. France 89(1961), 461-513.

103. Borel, A., Hirzebruch, F. Characteristic classes and homogeneous spaces I, Am. J. Math. 80(1958), 459-538; II, Am. J. Math. 82(1960), 491-504.

104. Borel, A., Narasimhan, R. Uniqueness conditions for certain holomorphic mappings. Invent. Math. 2(1967), 247-255.

105. Borel, A., Serre, J. P. Groupes de Lie et puissances reduites de Steenrod. Am. J. Math. 75(1953), 409-448.

106. Borel, A., Serre, J. P. Le theoreme de Riemann-Roch d'apres Grothendieck. Bull. Soc. Math. France 86(1958), 97-136.

107. Boothby, W. M. Homogeneous complex compact manifolds in differential geometry. Proc. Symp. Pure Math. III (1961), 144-154.

108. Bott, R. Homogeneous vector bundles. Ann. Math 66(1957) 203-248.

109. Bott, R. The space of loops on a Lie group. Mich. Math. J. 3(1958), 35-61.

110. Bott, R. Vector fields and characteristic classes. Mich. Math. J. 14(1967), 231-244.

111. Bott, R. On the topological obstructions to integrability. Proc. Symp. Pure Math. AMS 16(1970), 127-131.

112. Bott, R., Haefliger, A. On characteristic classes of $\Gamma$- foliations. Bull. Am. Math. Soc. 78(1972), 1039-1044.

113. Bourguignon, J. P. Premieres Formes de Chern des varietes Kahlerennes Compactes (d'apres E. Calabi, T. Aubin et S. T. Yau). Seminaire Bourbaki 507, Springer Lecture Notes 710(1979), 1-21.

114. Brendon, G. Introduction to Compact Transformation Groups. Academic Press, New York (1972).

115. Brenton, L., Morrow, J. Compactifying $C^n$. Symp. Pure Math. Vol. XXX, Part I (1977), 217-246.

116. Brieskorn, E. Ein satz uber die Komplexen Quadriken. Math. Ann. 155(1964), 187-193.

117. Brieskorn, E. Beispiele zur Differentialtopologie von singularitaten. Invent. Math. 2(1966), 1-14.

118. Brieskorn, E. Examples of singular normal complex spaces which

are topological manifolds. Proc. Nat. Acad. Sci. USA 55(1966), 1359-1397.

119. Brieskorn, E. — Rationale Singularitaten Komplexen Flachen. Invent. Math. 4(1968), 336-358.

120. Brieskorn, E., Van De Ven, A. — Some complex structures on product of homotopy spheres. Topology 7(1968), 389-393.

121. Brody, R. — Intrinsic metrics and measures on compact complex manifolds. Theses, Harvard (1975).

122. Brody, R., Green, M. — A family of smooth hyperbolic hypersurfaces in $P^3$. Duke Math. J. 44(1977), 873-874.

123. Browder, W. — Manifolds and Homotopy Theory, Manifolds. Amsterdam; Lecture notes 197, Springer Verlag (1970) 17-35.

124. Browder, W. — Poincare spaces, their normal fibrations and surgery. Invent. Math. 17(1972), 191-202.

125. Browder, W. — Surgery on Simply Connected Manifolds. Springer Verlag (1972).

126. Browder, F. (Editor) — Mathematical developments arising from Hilbert problems. Symp. Pure Math. Vol. XXVIII, Parts I and II, Am. Math. Soc. (1976).

127. Burniat, P. — Surfaces algebriques detee d'une faisceaux lineaire de courbes de genere deux: Hommage au Professeur Godeaux. Librairie Universitaire Louvain (1968), 103-111.

128. Burns, D. Jr., Rapport, M. — On the Torelli Theorem for Kahlerian K3 Surfaces. Ann. Sci. ENS (1975), 235-274.

129. Burns, D. Jr., Wahl, J. M. — Local contributions to global deformations of surfaces. Invent. Math. 26(1974), 67-88.

130. Buttin, C., Molino, P. — Theoreme general d'equivalence pour les pseudogroupes de Lie plats transitifs. J. Diff. Geom. 9(1972), 347-354.

131. Byrnes, C. I., Hurt, N. E. — On the moduli of linear dynamical systems. Studies in Mathematics Supplementary Studies No. 4 Academic Press (1978).

132. Cairns, S. S. — The manifold smoothing problem. Bull. Am. Math. Soc. 67(1961), 237-238.

133. Calabi, E. — The space of Kahler metrics. Proc. Int. Congr. Math. Amsterdam, Vol. 2(1954), 206-207.

134. Calabi, E. — On Kahler manifolds with vanishing canonical class,

Algebraic Geometry and Topology, Princeton Univ. Press (1957).

135. Calabi, E. Construction and properties of some 6 dimensional almost complex manifolds. Trans.Am.Math.Soc. 87(1958),407-483.

136. Calabi, E. On compact Riemann manifolds with constant curvature I, Proc.Symp.Pure Math. Vol.III,(1961), 155-180.

137. Calabi, E., Eckmann, B. A class of compact complex manifolds. Ann.Math. 58(1953),494-500.

138. Calabi, E., Rosenlicht, M. Complex analytic manifolds without countable basis. Proc.Amer.Math.Soc. (1953),335-340.

139. Calabi, E., Vesentini, E. On compact locally symmetric manifolds. Ann.Math. 71(1960),472-507.

140. Cartan, E. Les groupes de transformations continus infinis, simples. Ann.Sci. ENS (1909),93-161.

141. Cartan, E. Groupes infinis, system differentiels, theories d'equivalence (571-714), Le structure des groupes infinis (1335-1384), Oeuvres Completes: II Vol.2, Gauthier Villars, Paris (1953).

142. Cartan, H. Quotients of complex analytic spaces. Function Theory. Tata Inst. and Oxford Univ.Press (1960).

143. Cartan, H., Eilenberg, S. Homological Algebra. Princeton Univ.Press.

144. Cartan, H., Serre, J. P. Espaces fibres et groupes d'homotopic I: Construction generales. C.R.Acad.Sci. Paris 243(1952),288-290.

145. Cartan, H., Serre, J. P. Un theoreme de finitude concernant les varietes analytiques compactes. C.R.Acad.Sci. 237 (1953), 128-130.

146. Castelnuovo, G. Memorie scelte . Zanichelli, Bologna (1937).

147. Castelnuovo, G., Enriques, F. Die algebraischen Flachen von Gesichpunkte der birationalen transformationen aus, Enzyklop. der Math.Wiss. III 6B (1914),674-768.

148. Cathelineau, J. L. Deformations equivariantes de spaces analytiques complexes compacts. Ann.Scient. ENS(1978), 391-406.

149. Chen, B. Y. Surfaces admettang une metrique Kahlerian de Bochner. C.R.Acad. Paris 282(1976),643-644.

150. Chen, B. Y., Ogiue, T. Some characterisation of complex space forms in terms of Chern classes. Quart.J.Math.(Oxford) 26(1975),459-464.

151. Chern, S. S. Characteristic classes of Hermitian manifolds. Ann. Math. 47(1946), 85-121.

152. Chern, S. S. On the characteristic classes of complex sphere bundles and algebraic varieties. Am. J. Math. 75(1953), 565-597.

153. Chern, S. S. An elementary proof of the existence of isothermal coordinates on a surface. Proc. Am. Math. Soc. 6(1955), 771-782.

154. Chern, S. S. The geometry of G-structures. Bull. Am. Math. Soc. 72(1966), 167-219.

155. Chern, S. S. Complex Manifolds Without Potential Theory. Van Nostrand (1967).

156. Chern, S. S. Geometry of characteristic classes. Proc. 13th Biennial Seminar, Canadian Math. Cong. (1972).

157. Chern, S. S., Moser, J. Real hypersurfaces in complex manifolds. Acta Math. 133(1974), 219-271.

158. Chern, S. S., Simons, J. Characteristic forms and geometric invariants. Ann. Math. 99(1974), 48-69.

159. Chevalley, C, Introduction to the theory of algebraic functions of one variable. Am. Math. Soc. Surveys 6(1951).

160. Chevalley, C. Fundaments de la geometric algebriques. Secretarial Math. (1958).

161. Chow, W. L. On compact complex analytic varieties. Am. J. Math. 71(1949), 893-914.

162. Chow, W. L., Kodaira, K. An analytic surface with two independent meromorphic functions. Proc. Nat. Acad. Sci. USA 38(1952), 319-325.

163. Chow, W. L. On meromorphic maps of algebraic varieties. Ann. Math. 89(1969), 391-403.

164. Clemens, C. The Picard-Lefchetz theorem for families acquiring ordinary singularities. Trans. Am. Math. Soc. 139(1969), 93-108.

165. Clemens, C, Griffiths, P. The intermediate of Jacobian of the cubic threefold. Ann. Math. 96(1972), 281-356.

166. Cohen, M. M. A Course in Simple Homotopy Theory. Graduate Text in Math. 10, Springer Verlag (1978).

167. Commichau, M. Deformationen Kampakter Komplexer Mannigfaltig Keiten. Math. Ann. 213(1975), 43-96.

168. Conforto, F. Le Superier Razionali. Zanichelli, Bologna (1939).

169. Conforto, F. Abelsche Funktionen und Algebraische Geometrie. Springer Verlag (1956).

170. Conn, J.F. A new class of counter examples to the integrability problem. Proc. Nat. Acad. Sci. USA 74(1977), 2655-2658.

171. Conn, J.F. Non-abelian minimal closed ideals of transitive Lie algebras. Ph.D. Thesis, Princeton Univ. Press. Princeton, N.J. (1978).

172. Conner, P.E., Raymonds, F. Actions of compact Lie groups on aspherical manifolds; Topology of Manifolds, Proc. Univ. Georgia, Topology of Manifolds Inst. (1969), 227-264, Markham, Chicago 1971.

173. Conner, P.E., Raymonds, F. Manifolds with no periodic maps. Proc. 2nd Nat. Congr. on Compact Transformation Groups, Univ. Mass. Amherst (1971); Part II, Lecture Notes 299, Springer Verlag (1972), 81-101.

174. Conner, P.E., Raymonds, F, Realising finite groups of homomorphisms from homotopy classes of self-homotopy equivalences: Manifolds, Tokyo (1973), Tokyo Univ. Press (1975), 231-237.

175. Conte, A. Probleme di razionalita per la verieta algebriche a tre dimensioni. Bolletino UM.I. 5(1967).

176. Conte, A., Murre, J.P. On quartic threefolds with double line. I; II, Indag. Math. 39(1977), 145-175.

177. Cornalba, M. Two theorems on modifications of analytic spaces. Invent. Math. 20(1973), 227-248.

178. Cornalba, M. Complex tori and Jacobians, Complex Analysis and its Applications, Vol. I. IAEA (1977).

179. Cornalba, M, Griffiths, P. Some transcendental aspects of algebraic geometry. Proc. Symp. Pure Math. Vol. 29(1975), 3-110.

180. Courant, R., Hilbert, D. Methods of Mathematical Physics, Vol. I and II, Interscience, New York.

181. Cowsik, R.C., Nori, M.V. On Cohen -Maculay rings. J. Algebra 38(1976), 536-538.

182. De La Harpe, D. Introduction to complex tori, Complex Analysis and its Applications, Vol. I, IAEA (1977).

183. Deligne, P. Theorie de Hodge I, II, III, Proc. Int. Congr. Nice (1970), Publ. IHES, Bures-sur Yvette 40(1972), 5-58; 45(1974), 3-78.

184. Deligne, P. Le conjecture de Weil pour les surfaces $K_3$. Invent. Math. 15(1972), 206-226.

185. Deligne, P. Groupes de monodromic en geometric algebrique. Lecture Notes 340, Springer Verlag (1973).

186. Deligne, P., Griffiths, P., Morgan, J, Sullivan, D. On the homotopy type of compact Kahler manifolds. Invent. Math. 29 (1975), 245-274.

187. Deligne, P., Mumford, D. The irreducibility of the space of curves of given genus. Publ. IHES, Bures-sur-Yvette 36(1969), 75-110.

188. Deligne, P., Rapoport, M. Les schemes de modules de Courbes elliptiques. Lecture Notes 349, Springer Verlag (1973), 143-346.

189. Demazure, M., Grothendieck, A. Schemes en groupes. Sem. de Geom. Algebriques. Publ. IHES, (1963/64).

190. Diederich, K., Fornaess, J. E. Exhaustion of Functions and Stein Neighbourhoods. Proc. Nat. Acad. Sci. USA 72(9) (1975), 3279-3280.

191. Diederich, K., Fornaess, J. E. Bounded strictly plurisubharmonic functions. Invent. Math. 39(1977), 129-141.

192. Diederich, K., Fornaess, J. E. Pseudoconvex domains with real analytic boundaries. Ann. Math. 107(1978), 37-384.

193. Dieudonne, J. Algebraic Geometry. Advances in Math. 3(1969), 233-321.

194. Dieudonne, J. Fundaments de la Geometrie Algebrique Moderne. Advances in Math. 3(1969), 322-413.

195. Dieudonne, J., Grothendieck, A. Elements de geometric algebraique (Known as EGA ) Publ. IHES. (1960).

196. Dolbeault, P. Sur la cohomologie des varietes analytiques complexes. C. R. Acad. Sci. Paris 236(1953), 175-177.

197. Dolbeault, P. Formes differentielles et cohomologie sur une variete analytique complexes I, Ann. Math. 64(1956), 83-130; II, ibid. 65(1957), 282-350.

198. Dold, A. Erzengende der Thomshen Algebra. N. Math. Zeit. 65(1956), 25-35.

199. Dolgacev, I. On the purity of the degeneration loci of families of curves. Invent. Math. 8(1969), 34-54.

200. Donin, I. F. Conditions for the triviality of deformations of holomorphic fibre bundles over a compact complex space. Mat. Sb. (NS) 77(119)(1968), 602-623.

201. Donin, I. F. Conditions for the triviality of deformations of complex structures. Math. USSR Sb. 10(1970), 557-568.

202. Donin, I. F. Complete families of germs of complex spaces. Math. USSR Sb. 18(1972), 394-406.

203. Donnelly, H. Meenakshisundaram's coefficients on Kahler manifolds. Proc. Symp. Pure Math. Vol. 27(1973), 195-204.

204. Donnelly, H. Symmetric Einstein spaces and spectral geometry. Indiana Univ. Math. J. 24(1974), 603-606.

205. Donnelly, H. Topology of Einstein Kahler metric. J. Diff. Geom. 11(1976), 259-264.

206. Douady, A. Varietes et espaces mixtes, Deformations regulieres obstruction primaire a la deformation. Seminaire H. Cartan 13 Annee (1960/61) Expose 2, 3, 4.

207. Douady, A. Les probleme des modules pour sous espaces analytiques complexes (d'apres M. Kuranishi), Seminaire Bourbaki, Expose 277, (1964).

208. Douady, A. Le probleme de modules pur sous-espaces analytiques compacts d'un espaces analytique donne. Ann. Inst. Fourier (Grenoble) 16(1966), 1-95.

209. Douady, A. Les problemes des modules en geometrie analytique complexe. Proc. Int. Congr. Moscow (1966).

210. Douady, A. Flatness and privelege. L'enseignement Math. 15(1968), 45-74.

211. Douady, A. De probleme des modules locale pour les espaces C-analytique compacts. C. R. Acad. Sci. Paris. Ser. A-B 277(1973), A 901-A 904.

212. Douady, A. Le probleme des modules locaux pour les espaces C-analytiques compacts. Ann. Sci. Ecole Nor. Sup. IV 4(1974), 567-602.

213. Duistermatt, J. J., Hormander, L. Fourier integral operators II. Acta Math. 128(1972), 183-269.

214. Durfee, A. H. Foliations of odd dimensional spheres. Ann. Math. 96(1972), 407-411, Eratum. Ann. Math. 97(1973), 187.

215. Earle, C. J, Eells, J. Deformations of Riemann Surfaces, Modern Analysis and Applications I. Lecture Notes 103, Springer Verlag (1969).

216. Eckmann, B. Complex analytic manifolds. Proc. Int. Congr. Math. Camb. Mass. 2(1950), 420-427.

217. Eckmann, B., Frolicher, A. Sur l'integrabilite de structures presque complexe. C.R.Acad. Sci. Paris 232(1951), 2284-2286.

218. Edge, W. L. A new look at the Kummar surfaces. Canad. J. Math. 19(1967), 952-967.

219. Edwards, R.D. The double suspension of a certain homology 3-sphere in $S^5$. Notices of Am. Math. Soc. 22(1975), A-334.

220. Edwards, R.D. The double suspension of PL-homology n-sphere. Proc. Topology Conf. Univ. Georgia (1975).

221. Eells, J. On the moduli of closed Riemann surfaces with boundary. Ann. Math. Studies. Princeton Univ. Press 86(1971), 119-130.

222. Eells, J., Kuiper, N.H. Manifolds which are like projective planes. Publ. IHES 11(1962), 1-46.

223. Eells, J., Kuiper, N.H. An invariant for certain smooth manifolds. Annali de Mat. pure et appl. 60(1963), 93-100.

224. Eells, J., Lemaire, L. A report on harmonic maps. Bull. Lond Math Soc. 10(1978), 1-68.

225. Eells, J, Sampson, J.H. Harmonic mappings of Riemannian manifolds. Am. J. Math. 86(1964), 109-160.

226. Eells, J., Sampson, J.H. Variational theory in fibre bundles. Proc. YS-Japan Seminar on Diff. Geom. Kyoto (1965), 22-33.

227. Ehresmann, C. Sur les varietes presque complexe. Proc. Int. Congr. Math. Cambridge (1950).

228. Ehresmann, C. Sur la theorie des varietes feuilletees. Prend. Mat. e. Appl. Serie V, Vol. X(1951), 1-19.

229. Ehresmann, C. Introduction a la theorie des structures infinitesimales et des pseudogroupes de Lie; Colloque de Geometrie Differentialle de Strasbourg, Centri National de la Recherche Scientific, Paris (1953), 97-110.

230. Eilenberg, S. Cohomology and continuous maps. Ann. Math. (2), 41(1940), 231-251.

231. Eilenberg, S. Singular homology theory. Ann. Math. (2), 45(1944), 407-447.

232. Eilenberg, S., Maclane, S. Relation between homology and homology groups of spaces. Ann. Math. (2) 45(1945), 480-509.

233. Eilenberg, S., Maclane, S, On the groups $H(\pi, n)$. Ann. Math. (2) 62(1954), 513-517.

234. Eilenberg, S., Steenrod, N. Foundations of Algebraic Topology. Princeton Univ. Press, Princeton (1952).

235. Enriques, F. Le superficie algebraiche. Zanichelli, Bologna.

236. Ephraim, R. Cartesian product structure of singularities. Proc. Symp. Pure Math. AMS 30(1)(1977), 21-23.

237. Fano, G. Sulle verieta algebriche a tre diensioni a curvesezioni canoniche. Med. Accad. It. 8(1937), 23-64.

238. Farrell, F. T., Hsiang, W. C. h-cobordant manifolds are not necessarily homeomorphic. Bull Am. Math. Soc. 73(1967), 741-744.

239. Fefferman, C. The Bergman kernel and biholomorphic mappings of pseudo convex domains. Invent. Math. 26(1974), 1-65.

240. Fischer, G. Projective embetting Komplexer Mannigfaltigkeiten. Math. Ann. 234(1978), 45-50.

241. Fisher, G., Grauert, H. Lokal-triviale Familien Kompakter Komplexer Mannigfaltigkeiten. Nachr. Akad. Gottingen Math. Phys. 2(1965), 89-94.

242. Fogarty, J. Algebraic families on an algebraic surface. Am. J. Math. (1968), 511-521.

243. Forness, J. E. An increasing sequence of Stein manifolds whose limit is not Stein. Math. Ann. 223(1976), 275-277.

244. Forster, O. Plongements des varietes de Stein. Comm. Math. Helv. 45(1970), 170-184.

245. Forster, O. Power series methods in deformation theory. Proc. Symp. Pure Math. AMS 30(2)(1977), 199-217.

246. Forster, O., Knorr, K. Ein neuer Beweis des Satzes von Kodaira-Nirenberg-Spencer. Math. Z. 139(1974), 257-291.

247. Forster, O., Knorr, K. Uber die Deformationen von Vektorraumbundeln auf Kompakten Komplexen Raumen. Math. Ann. 209(1974), 29-34.

248. Forster, O., Knorr, K. Konstrucktion verseller Familien Kompakter Komplexer Raume. Lecture Notes 705, Springer Verlag (1979).

249. Fort, K. M. Jr. (Editor) Topology of 3-manifolds and related topics. Prentice Hall (1962).

250. Frankel, T. Manifolds with positive curvature. Pac. J. Math. ii(1961), 165-174.

251. Freyd, P. Abelian Categories. Harper and Row (1964).

252. Frish, J. Note Zu der Arbeit von J. Frish: Points de platitude d'un morphisme d'espaces analytiques. Invent. Math. 4(1967), 118-138.

253. Frohlicher,A. Zur Differentialgeometrie der complexen structuren. Math.Ann. 129(1955),50-95.

254. Frohlicher,A. Relations between the cohomology groups of Dolbeault and topological invariants. Pro.Nat.Acad.Sci.USA 41(1955),641-644.

255. Frohlicher,A.,Kobayashi,S., Nijenhuis,A. Deformation theory of complex manifolds. Techn.Rep.10 Univ.Washington (1960).

256. Frohlicher,A.,Nijenhuis,A. Theory of vector valued differential forms. Part I. Proc.Kon.Ned.Akad.Wel.Amsterdam 59(1956),338-359; Part II, ibid,61(1958),414-429.

257. Frohlicher,A.,Nijenhuis,A. Some new cohomology invarianst for complex manifolds. Proc.Kon.Ned.Akad.Wel.Amsterdam 540-564.

258. Frohlicher,A.,Nijenhuis,A. A theorem on the stability of the complex structures. Proc.Nat.Acad.Sci.USA 43(1957),229-241.

259. Fujita,U. Sur les familes d'ensablles analytiques. J.Math.Soc. Japan 16(1964),379-405.

260. Fuks,D.B. Characteristic classes of foliations. Russian Math. Survey. 28(1973),1-16.

261. Fulton,W. Algebraic Curves. Benjamin,New York (1969).

262. Fulton,W. Ample vector bundles,Chern classes and numerical criteria.
32(1976,171-178.

263. Gabriel,P. Des categories abeliennes. Bull.Soc.Math.France 90(1962),323-348.

264. Galewski,D.E.,Stern,R.J. Classification of simplical triangulations of topological manifolds. Bull.Am.Math. Soc. 82(1976),916-918.

265. Galligo,A,Hauzel,C. Moduli des singularites d'apres Verdier et Grauert, Singularities Cargase (1972),139-163,Astriques 7 et 8 Pub.Soc.Math.France (1973).

266. Ganesh,M. Projective structures on compact complex manifolds: Geometric Theory I, J.Indian Math.Soc. 42(1978), 67-84; II, Publ.Ramanujan Inst.,Madras-5,4(1979), 165-179.

267. Gelfand,I.M. Cohomologies of the Lie algebra of formal vector fields. Math.USSR Izv.4(1970),327-342.

268. Gelfand,I.M.,Fuks,D.B. Cohomologies of the Lie algebra of tangent vector fields of smooth manifolds I,Functional analysis and applications 3(1969),194-210;ibid 4(1970),110-116.

269. Gerstenhaber, M. On the deformation of rings and algebras I, Ann. Math. 79(1964), 59-104; II, ibid 84(1966), 1-16; III, ibid 88(1968), 1-34.

270. Giesekar, D. On the moduli of vector bundles over an algebraic surface, Ann. Math. 106(1977), 45-60.

271. Giesekar, D. Global moduli for surfaces of general type. Invent. Math. 43(1977), 232-282.

272. Gilkey, P. Curvature of the eigenvalues of the Laplacian for geometrical elliptical complexes. Ph.D. Thesis Harvard Univ. USA (1972).

273. Gilkey, P. Spectral geometry of real and complex manifolds. Proc. Symp. Pure Math. AMS 27(1975), 265-280.

274. Gilkey, P. Spectral geometry and the Kahler condition for complex manifolds. Invent. Math. 26(1974), 231-238.

275. Gilkey, P. The spectral geometry of Riemannian manifold. J. Diff. Geom. 10(1975), 601-618.

276. Gilkey, P., Sacks, J. Spectral geometry and manifolds of constant holomorphic sectional curvature. Proc. Symp. Pure Math. AMS 27(1975), 281-285.

277. Godbillon, C., Vey, J. Un invariant des fecilletages de Codimension un. C.R. Acad. Sci. Paris 273(1971), 92-95.

278. Godeaux, L. Sur le systeme canonique des surfaces algebriques triples. Acad. Roy. Belg. Bull. Sci. (5) 56(1970), 856-864, (5) 56(1970), 1016-1023.

279. Godeaux, L. Sur les surfaces algebriques de genre zero et de genre lineaire un. Acad. Roy. Belg. Bull. Sci. (5) 56(1970), 560-564 and 649-653.

280. Godeaux, L. Nouvelles recerches sur les involutions cycliques apparitienant a une surface I, II, III, Acad. Roy. Belg. Bull. Sci. (5) 56(1970), 1192-1204.

281. Godement, R. Topologie Algebrique et Theorie des Faisceaux. Acta Sci. et Ind. 1252, Hermann (1964).

282 Goldberg, S. I., Kobayashi, S. On holomorphic bisectional curvature. J. Diff. Geom. 1(1967), 225-233.

283. Goldschmidt, H. Le troisieme theoreme fundamental. J. Diff. Geom. 6(1972), 357-373.

284. Goldschmidt, H. Equations formallen et transitive. J. Diff. Geom. 7(1972), 67-95.

285. Goldschmidt, H. Le cohomologie de Spencer. J. Diff. Geom. 11(1976), 167-223.

286. Goldschmidt, H. The integrability problem for Lie equations. Bull. Am. Math. Soc. 84(1978), 531-546.

287. Goldschmidt, H., Spencer, D. C. On the non-linear cohomology of Lie equations I, II, Acta Math. (1976), 103-239.

288. Goldschmidt, H., Spencer, D. C. On the non-linear cohomology of Lie equations III, IV, J. Diff. Geom. 13(1978).

289. Goodmann, J. E. Affine open subsets of algebraic varieties and ample divisors Ph. D. Thesis, Columbia Univ. (1967).

290. Goren, R. Characterisation and algebraic deformation of projective space. J. Math. Kyoto Univ. 8(1968), 41-47.

291. Grabowski, J. Isomorphisms and ideals of the Lie algebra of vector fields. Invent. Math. 50(1978), 13-33.

292. Grauert, H. On Levi's problem and the embedding of real analytic manifolds. Ann. Math. 68(1958), 460-472.

293. Grauert, H. Ein Theorem der analytischen Garbentharie und die Modulraume Komplexen Strukturen. Publ. IHES, Bures-sur Yvette 5(1960).

294. Grauert, H. On the number of moduli of complex structures, Function Theory, Tata Inst. Oxford Univ. Press (1960).

295. Grauert, H. Uber Modifikationen und exzeptionelle analytische Mergen. Math. Ann. 146(1962), 331-368.

296. Grauert, H. Uber die Deformation isolierter singularitaten Mergen. Invent. Math. 15(1972), 171-198.

297. Grauert, H. Der Satz von Kuranishi fur Kompakte Komplexe Raume. Invent. Math. 25(1974), 107-142.

298. Grauert, H. Uber die Deformation von Pseudo gruppen Strukturen. Astriques 32, 33, Soc. Math. France (1976), 141-150.

299. Grauert, H., Remmert, R. Komplex Raume. Math. Ann. 136(1958), 245-328.

300. Grauert, H., Remmert, R. Theorie der Steinschen Raume. Springer Verlag, (1977).

301. Grauert, H., Remmert, R., Riemanschneider, O. Analytische Stellenalgebren, Springer (1971).

302. Gray, J. W. Some global properties of contact structures. Ann. Math. 69(1959), 421-450.

303. Greb, W., Halperin, S., Vanstone, R. Connections, Curvature and Homology. Academic Press (1972).

304. Green, R. E., Wu, H. Curvature and complex analysis I, Bull. Am. Math. Soc. 77(1971), 1045-1049, II, ibid 78(1972), 866-870,

III, ibid 79(1973), 606-608.

305. Green, R. E., Wu, H. A theorem in complex geometric function theory, Value distribution theory, Part A, Marcel Dekker, New York (1974), 145-167.

306. Green, R. E., Wu, H. Some function theoretic properties of non-compact Kahler manifolds. Proc. Symp. Pure Math. Vol. 27, Part II, AMS (1975), 33-41.

307. Green, R. E., Wu, H. Analysis on noncompact Kahler manifolds. Proc. Symp. Pure Math. Vol. 30, Part II, AMS (1977), 69-100.

308. Greiner, P. C., Stein, E. M. Estimates for the $\bar{\partial}$ Neumann problem. Math. Notes, Princeton Univ. Press (1977).

309. Griffiths, P. A. Some geometric and analytic properties of homogeneous complex manifolds. Acta Math. 110(1963), 115-208.

310. Griffiths, P. A. On the existence of a locally complete germ of deformation of certain G-structures. Math. Ann. 159(1966), 151-171.

311. Griffiths, P. A. The extension problem in complex analysis I, Proc. Conf. Complex Analysis, Minneapolis, Springer Verlag (1966).

312. Griffiths, P. A. Some remarks and examples of continuous systems and moduli. J. Math. Mech. 16(1967), 789-802.

313. Griffiths, P. A. Periods of integrals on algebraic manifolds, I, II, Am. J. Math. 90(1968), 586-626; 805-865; III Publ. IHES, Bures-sur-Yvette 38(1970).

314. Griffiths, P. A. On periods of certain rational integrals. I, II, Ann. Math. 90(1969), 460-495, 498-541.

315. Griffiths, P. A. Periods of integrals on algebraic manifolds: Summary of main results and discussion of open problem. Bull. Am. Math. Soc. 76(1970), 228-296.

316. Griffiths, P. A. Deformation of Complex Structures, Global Analysis, AMS (1970).

317. Griffiths, P. A. Differential Geometry and Complex Analysis. Proc. Symp. Pure Math. 27(1975), 43-64.

318. Griffiths, P. A., Adams, J. Topics in Analytic and Algebraic Geometry. Princeton Univ. Press (1974).

319. Griffiths, P. A., Harris, J. Principles of Algebraic Geometry, John Wiley (1978).

320. Griffiths, P. A., Schmidt, W. Recent developments on Hodge theory. Proc. Int. Col. Discrete Subgroups of Lie Groups and Applications to Moduli, Tata Inst. Bombay (1973).

321. Griffiths, P.A., Schmidt, W. Locally homogeneous complex manifolds, Acta Math. 123(1970), 253-302.

322. Grothendieck, A. Theoreme de dualite pour faisceaux algebrique coherents, Sem. Bourbaki 149(1956/57).

323. Grothendieck, A. Sur la classification des fibres holomorphes sur la sphere de Riemann, Am. J. Math. 79(1957), 121-138.

324. Grothendieck, A. Sur quelques points d'algebre homologique, Tohoku Math. J. (2) (1957), 119-221.

325. Grothendieck, A. La theorie des classes de Chern, Bull. Soc. Math. France 86(1958), 137-154.

326. Grothendieck, A. Quelques problemes des modules, Sem. Henri Cartan 1960-1961, Expose 16.

327. Grothendieck, A. Construction de l'espace de Teichmuller, Sem. Henri Cartan 13 Annee (1960/61), Expose 17, Tata Inst. Oxford Univ. Press (1975).

328. Grothendieck, A. Les schemes de Hilbert, Sem. Bourbaki, 221(1960/61).

329. Grothendieck, A. Technique de construction en geometrie analytique I, a X, Cartan Seminar, 1960/61, Secretariat Mathematique, Paris (1962).

330. Grothendieck, A. On the de Rham cohomology of algebraic varieties, Publ. Math. IHES 29(1966), 351-359.

331. Guillemin, V.W. The integrability problem for G-structures, Trans. Am. Math. Soc. 116(1965), 544-560.

332. Guillemin, V.W. A Jordan Holder decomposition for a certain class of infinite dimensional Lie algebras, J. Diff. Geom. 2(1968), 313-345.

333. Guillemin, V.W. Cohomologies of vector fields on a manifold. Advances in Math. 10(1973), 192-220.

334. Guillemin, V.W., Steinberg, S. An algebraic model of transitive differential geometry, Bull. Am. Math. Soc. 70(1964), 16-47.

335. Guillemin, V.W., Steinberg, S. Deformation theory of pseudogroup structures, Memoirs AMS 64(1966).

336. Guillemin, V.W., Steinberg, S. The Lewy counter example and local equivalence problem for G-structures, J. Diff. Geom. 1 (1967), 127-131.

337. Gunning, R.C. Lectures on Riemann Surfaces, Princeton Univ. Press (1965).

338. Gunning, R.C. On uniformization of complex manifolds - The role of connection, Lecture Notes, Princeton Univ. Press (1979).

339. Gunning,R.C.,Rossi,H. Analytic Functions of Several Complex Variables, Prentice Hall (1965).

340. Gunning,R.C.,Narasimhan,R. Immersions of open Riemann surfaces, Math. Ann. 174(1967),103-108.

341. Haboush,W.J. Reductive groups are geometrically reductive, Am. Math. 102(1975),67-83.

342. Haefliger,A. Structures feuilletees et cohomologie a valuer dans un faisceau de groupoids, Comm.Math.Helv. 32(1958),248-329.

343. Haefliger,A. Plongements differentiables de varietes dans varietes, Comm.Math.Helv. 36(1961),47-82.

344. Haefliger,A. Varietes feuilletes, Ann.L'ecole Norm.Sup. Paris 16(1962),367-379.

345. Haefliger,A. Homotopy and integrability,Manifolds- Amsterdam (1970); Lecture Notes 197,Springer Verlag (1971), 133-163.

346. Hamilton,R.S. Deformation of complex structures on pseudoconvex domains, Cornell Univ.(Preprint) (1972).

347. Hamilton,R.S. Deformation of complex structures on manifolds with boundary I,II,III,IV, Cornell University (USA) Preprint.

348. Hamilton,R.S. Harmonic maps on manifolds with boundary, Lecture Notes 471,Springer Verlag (1975).

349. Hammond,W.F. The Hilbert modular surface of a real quadratic field, Math.Ann. 200(1973),25-45.

350. Hangen,T. Sur l'integrabilite de certain G-structures tensor-ielles, Comptes Rendes de Sciences d L'Academie des Sciences,Paris (1979),835-837.

351. Harder,G.,Narasimhan,M.S. On the cohomology of moduli spaces of vector bundles, Math.Ann.212(1975),215-248.

352. Harrison,D.K. Commutative algebras and cohomology, Trans. Am. Math.Soc. 104(1962),191-204.

353. Hartshorne,R. Residues and Duality, Lecture Notes 20,Springer Verlag (1966).

354. Hartshorne,R. Ample Subvarieties of Algebraic Varieties, Lecture Notes 156,Springer Verlag (1970).

355. Hartshorne,R. On the DeRham cohomology of algebraic varieties, Publ.IHES 45(1975),5-99.

356. Hartshorne,R. Algebraic Geometry, Graduate Text 52,Springer Verlag (1977).

357. Heaps,T. Almost complex structures on 8 and 10 dimensional manifolds,Topology 9(1970),111-119.

358. Hecke,E. Hohere Modulfunktionen und ihre Anwendungen auf die Zahlentheorie, Math.Ann.71(1972),1-37.

359. Helgason,S. Differential Geometry and Symmetric Spaces, Academic Press, New York (1962).

360. Hempel,J. 3-manifolds, Ann.Math.Studies 86, Princeton Univ. Press (1976).

361. Higuchi,T.,Tsuboi,M. On the Kodaira-Spencer sequence of complex spaces, Math.Z. 116(1970),331-337.

362. Hilbert,D.,Cohn Vossen,S. Geometry and the Immagination,Chelsea Publ.Co. New York (1972).

363. Hironaka,H. A note on algebraic geometry over ground rings, Ill. J.Math. 2(1958),355-366.

364. Hironaka,H. An example of a nonKahlerian complex analytic deformation of Kahlerian complex structures, Ann. Math. 75(1962),642-648.

365. Hironaka,H. Resolution of singularities of an algebraic variety over a field of characteristic zero, Ann.Math. 79(1964),109-306.

366. Hironaka,H. Desingularization of complex analytic varieties, Proc.Int.Congr.Math.Nice (1970),Vol.2, 627-631.

367. Hironaka,H. Bimeromorphic smoothing of a complex space, Math. Inst. Warwick Univ. Preprint (1971).

368. Hironaka,H. Flattening of Analytic Maps,Manifolds, Tokyo (1973), Iwanami Shoten (1975),313-321.

369. Hironaka,H. Flattening theorem in complex analytic geometry, Am.J.Math. 97(1975),503-537.

370. Hironaka,H.,Matsumara,H. Formal functions and formal imbeddings, J.Math.Soc.Japan 20(1968),52-82.

371. Hirsh,M.W. On embedding of differentiable manifolds in Euclidean spaces, Ann.Math. 67(1958),566-571.

372. Hirsh,M.W. Obstruction theorems for smoothing manifolds and maps, Bull.Am.Math.Soc.69(1963),352-356.

373. Hirsh,M.W.,Mazur,B. Smoothing piecewise linear manifolds, Ann.Math. Studies 80, Princeton Univ.Press (1974).

374. Hirzebruch, F. Uber eine Klasse von einfachzusammenhangenden Komplexen Mannigfaltigkeiten, Math. Ann. 124(1951), 77-86.

375. Hirzebruch, F. Some problems on differentiable and complex manifolds, Ann. Math. 60(1954), 231-236.

376. Hirzebruch, F. Arithmetic genera and the theorem of Riemann-Roch for algebraic varieties, Proc. Nat. Acad. Sci. USA 40(1954), 110-114.

377. Hirzebruch, F. Komplexe Mannigfaltigkeiten, Proc. Int. Symp. Mathematics (1958), 119-136.

378. Hirzebruch, F. The topology of normal singularities of an algebraic surface ( d'apres Mumford), Seminaire Bourbaki, 250, (1961-62).

379. Hirzebruch, F. Topological Methods in Algebraic Geometry, Springer Verlag (1966).

380. Hirzebruch, F. Singularities and exotic spheres, Seminaire Bourbaki 314(1966-67).

381. Hirzebruch, F. The singularities of ramified coverings, Global Analysis (Kodaira Volume), Princeton Univ. Press (1969), 253-265.

382. Hirzebruch, F. The signature theorem: Reminiscences and Recreation, Prospects in Mathematics, Ann. Math. Studies, Princeton Univ. Press (1971).

383. Hirzebruch, F. Hilbert modular surfaces, L'Enseignement Math. 19(1972), 183-281.

384. Hirzebruch, F. Modulflachen und modulkurven zur symmetrischen Hilbertschen Modulgruppe, Ann. Scient. Ec. Norm. Sup. 11(1978), 101-166.

385. Hirzebruch, F., Hopf, H. Felder Von Flachenelementen in 4 dimensionelen mannigfaltigkeiten, Math. Ann. 136(1958), 156-172.

386. Hirzebruch, F., Kodaira, K. On complex projective spaces, J. Math. Pure Appl. 36(1957), 201-216.

387. Hirzebruch, F., Mayer, K. H. O(n)-Mannigflatigkeiten-exotische spheren und singulariton - Lecture Notes 57, Springer Verlag (1968).

388. Hirzebruch, F., Neumann, W. D., Koh, S. S. Differentiable Manifolds and Quadratic Forms, Marcel Dekker Inc., New York (1971).

389. Hirzebruch, F., Van De Ven, A. Hilbert modular surfaces and classification of algebraic surfaces, Invent. Math. 23(1974), 1-29.

390. Hirzebruch.F.,Zagier,D. Classification of Hilbert modular surfaces: Comples Analysis and Algebraic Geometry In Honour of Kodaira, Iwanami Shoten (1977),137-150.

391. Hitchin,N. Compact 4 dimensional Einstein 4-manifolds, J.Diff.Geom. 9(1974),435-441.

392. Hitchin,N. On the curvature of rational surfaces, Proc.Symp. Pure Math. 27(1975),65-80.

393. Hodge,W.V.D. The topological invariants of algebraic varieties, Proc.Int.Congr. Vol. 1(1950),182-191.

394. Hodge,W.V.D. Some recent results in the theory of algebraic varieties,J. London Math.Soc. 25(1950),143-157.

395. Hodge,W.V.D. The Theory and Application of Harmonic Integrals, Camb.Univ.Press (1952).

396. Hodgson,J.P.E. Subcomplexes of Poincare complexes, Bull.Am.Math. Soc. 80(1974),1146-1150.

397. Holm,P. (Editor) Real and Complex Singularities,Oslo (1976), Sijthoff and Noordhoff International Publishers.

398. Holmann,H. Quotienteuraume Komplexer Mannigfaltigkeiten nach Komplexen lieschen Automorphismen gruppen, Math.Ann. 139(1960),383-402.

399. Holmann,H. Quotienten Komplexer Raume, Math.Ann.142(1960/61), 407-440.

400. Hopf,H. Zur Topologie der Komplexen Mannigfaltigkeiten, Courant Volume,Interscience (1948),167-185.

401. Hopf,H. Schlitche Abbildungen und lokale Modificationen 4-dimensionalen komplexer Mannigfaltigkeiten, Comm.Math.Helv.29(1965),132-156.

402. Horikawa,E. On deformation of holomorphic maps I, J.Math.Soc. Japan 25(1973),372-396; II,ibid 26(1974),647-667; III,Math.Ann.222(1976),275-282.

403. Horikawa,E. On deformations of quintic surfaces, Invent.Math. 31(1975),43-85.

404. Horikawa,E. On the number of moduli of certain algebraic surfaces of general type, J.Fac.Sci.Univ.Tokyo 22(1975), 67-78.

405. Horikawa,E. On algebraic surfaces of general type with small $C_1^2$, I,Ann.Math.104(1976),357-386;II,Invent.Math. 37(1976),121-155; III,ibid 47(1978),209-248; IV, ibid 50(1979),103-128.

406. Horikawa, E. Tricanonical maps of numerical Godeaux surfaces, Invent. Math. 34(1976), 99-111.

407. Horikawa, E. On numerical Campedelli surfaces, Complex Analysis and Algebraic Geometry (in honour of Kodaira), Cambridge Univ. Press (1977), 113-118.

408. Horikawa, E. On the periods of Enriques surfaces, I, II, Proc. Japan Acad 53(1977), 124-127.

409. Hormander, L. Linear Partial Differential Operators, Springer Verlag (1963).

410. Hormander, L. An Introduction to Complex Analysis in Several Variables, Van Nostrand (1966).

411. Hormander, L. The spectral function of an elliptic operator, Acta Math. 121(1968), 193-218.

412. Hormander, L. Fourier Integral Operators I, Acta Math. 127(1971), 79-183.

413. Houzel, Ch. Geometrie analytique locale I a IV Seminaire H. Cartan 1960/61, Ecole Norm. Sup. Paris.

414. Howard, A., Smyth, B. On Kahler surfaces of nonnegative curvature, J. Diff. Geom. 5(1971), 491-502.

415. Hsiang, W. Y. Cohomology theory of topological transformation groups, Ergebisse der Mathematik und ihrer Grenzgebeite, 85, Springer Verlag (1975).

416. Hudson, J. F. P. Piecewise Linear Topology, Pergamon, New York (1969).

417. Huikeshoven, F. On the versal deformation of rational double points, Invent. Math. 20(1973), 15-34.

418. Hurewicz, W., Wallman, H. Dimension Theory, Princeton Univ. Press (1946).

419. Husemuller, D. Fibre Bundles, McGraw Hill, New York (1966).

420. Igusa, J. Fibre systems of Jacobian varieties I, Am. J. Math. 78(1956), 171-199; II, ibid 745-760; III, ibid 81(1945), 453-476.

421. Igusa, J. Arithmetic variety of moduli for curves of genus 2, Ann. Math. (2) 72(1960), 612-649.

422. Igusa, J. Betti and Picard numbers of abstract algebraic surfaces, Proc. Nat. Acad. Sci. USA 46(1960), 724-726.

423. Iitaka, S. Deformation of compact complex surfaces I, Global Analysis, Princeton Univ. Press and Tokyo Univ. Press (1969), 267-272; II, J. Math. Soc. Japan 22(1970), 247-261, III, ibid 23(1971), 692-705.

424. Iitaka, S. On D-dimension of algebraic varieties, J. Math. Soc. Japan 23(1971), 356-373.

425. Iitaka, S. On some new birational invariants of algebraic varieties and their applications to rationality of certain algebraic variety of dimension 3, J. Math. Soc. Japan 24(1972), 384-396.

426. Iitaka, S. Logarithmic forms of algebraic varieties, J. Fac. Sci. Univ. Tokyo 23(1976), 523-544.

427. Illman, S. Equivariant singular homology and cohomology for actions of compact Lie groups, Ph.D. Thesis, Princeton (1971).

428. Illman, S. Equivariant singular homology and cohomology for sections of compact Lie groups, Proc. Second Conference on compact transformation groups, Part I, Lecture Notes 298(1972), 403-415.

429. Illusie, L. Complexe cotangent et deformations I, II, Lecture Notes 239, 283 Springer Verlag (1971, 1972).

430. Inoue, M. On surfaces of class VII, Manifolds, Tokyo (1973); Invent. Math. 24(1974), 269-310; Hattori (Editor), Tokyo Univ. Press (1977), 389-392.

431. Inoue, M. New surfaces with no meromorphic functions I, II, Proc. Int. Congr. Math. Vol. I, Vancouver (1974), 423-426; Complex Analysis and Algebraic Geometry (Kodaira Volume) Edited by Baily and Shioda, Iwanami Shoten and Camb. Univ. Press (1977), 91-106.

432. Iskovski, V. A. Fano threefolds, Izv. AN USSR, Ser. Malen 41(1977), 516-562.

433. Iskovski, V. A., Manin, J. U. Three-dimensional quartics and counter examples to the Luroth problem, Math. USSR Sb. 15(1971), 141-166.

434. James, I. M. On sphere bundles, Bull. Am. Math. Soc. 75(1969), 617-621.

435. James, I. M. Bundles with special structures, Ann. Math. 89(1969), 359-360.

436. James, I. M., Thomas, E. Submersions and immersions of manifolds, Invent. Math. 2(1967), 171-177.

437. Johnson, F. E. Manifolds of homotopy type $K(\pi, 1)$, Proc. Camb. Phil. Soc. 70(1971), 387-393.

438. Kac, M. Can one hear the shape of a drum, Amer. Math. Monthly 73(1966), 1-23.

439. Kamber, F.W., Tondeur, Ph. Foliated bundles and characteristic classes, Springer Lecture Notes 493, (1975).

440. Kampen, E.R. Van On characterization of 2-dimensional manifolds, Duke Math. J. 1 (1935), 74-93.

441. Kas, A. On obstructions to deformations of analytic surfaces, Proc. Nat. Acad. Sci. USA 58(1967), 402-404.

442. Kas, A. On deformations of certain type of irregular algebraic surfaces, Am. Math. 90(1968), 789-804.

443. Kas, A., Schlessinger, M. On the versal deformation of a complex space with isolated singularity, Math. Ann. 196(1972), 23-29.

444. Kato, M. Complex structures on $S^1 \times S^5$, Proc. Jap. Acad. 49(1973), 575-577.

445. Kato, M. Topology of Hopf surfaces, J. Math. Soc. Jap. 27(1975), 222-238.

446. Kato, M. Complex manifolds containing global spherical shells, Proc. Int. Symp. Algebraic Geom. Kyoto (1977).

447. Katz, M. On the differential equations satisfied by period matrices, Publ. IHES, 35(1968), 71-106.

448. Kaup, B. Ein Kriterium fur platte holomorphe Abbildungen, Bayer Akad. Wiss. Math. Naturen. KL S-B(1969), 101-105.

449. Kavie, Y. Cohomologies of Lie algebras of vector fields with coefficients in adjoint representations, Publ. RIMS, Kyoto Univ. 14(1978), 487-509.

450. Kawai, S. On compact complex analytic manifolds of complex dimension 3, I, J. Math. Soc. Jap. 17(1965), 438-442, II, ibid 21(1969), 604-616.

451. Kawamata, Y. On deformations of compactifiable complex manifolds, Math. Ann. 235(1978), 247-265.

452. Kazdan, J. A remark on the preceeding paper of Yau, Comm. in Pure and Appl. Math. XXXI (1978), 413-414.

453. Kerner, H. Zur Theorie der deformationen Komplexer Raume, Math. Z. 103(1968), 389-398.

454. Kerner, H. Familien Komplexer Raume Zu gegebenen infinitesimalen Deformationen, Manus. Math. 1(1969), 317-337.

455. Kervaire, M. Non-parallelisability of the n-sphere $n > 7$, Proc. Nat. Acad. Sci. USA 44(1958), 280-283.

456. Kervaire, M. A manifold which does not admit any differential structure, Comm. Math. Helv. 34(1960), 257-270.

457. Kervaire, M., Milnor, J. Groups of homotopy spheres I, Ann. Math. 77(1963), 504-537.

458. Kiehl, R. Points de platitude d'un morphisme d'espaces analytiques, Invent. Math. 4(1967), 139-141.

459. Kiehl, R., Verdier, J. L. Ein einfacher Deweis des Koharenzsatzes von Grauert, Math. Ann. 195(1971/72), 24-50.

460. Kim, K. Deformations, related deformations and a universal subfamily, Trans. Am. Math. Soc. 121(1966), 505-515.

461. Kiremidjian, G. K. Deformations of complex structures on certain noncompact manifolds, Ann. Math. (2), 98(1973), 411-426.

462. Kiremidjian, G. K. On complex structures with a fixed induced C.R. structure, Ann. Math. 109(1979), 87-119.

463. Kirky, C., Siebermann, L. C. Foundational essays on topological manifolds, Smoothings and Triangulations, Ann. Math. Studies 88, Princeton Univ. Press (1977).

464. Kister, J. Microbundles and fibre bundles, Am. Math. 80(1964), 190-199.

465. Kleiman, S. L. Towards a numerical theory of ampleness, Ann. of Math. 84(1966), 293-344.

466. Kleiman, S. L. Completeness of characteristic systems, Adv. Math. 11(1973), 304-310.

467. Klingenberg, W. Manifolds with restricted conjugate locus, Am. Math. 78(1963), 525-547.

468. Knorr, K. Uber den Grauertschen Koharenzsatrz bei eigenlichen holomorphe Abbildungen I, II, Ann. Sci. Ecole Nor. Sup. Pisa 22(1968), 729-768; ibid 23(1969), 1-74.

469. Knorr, K., Schneider, M. Relativexzeptionelle analytische Mengen, Math. Ann. 193(1971), 238-254.

470. Knutsson, D. Algebraic Spaces, Lecture Notes 203, Springer Verlag (1971).

471. Kobayashi, S. Topology of positively pinched Kahler manifolds, Tohoku Math. J. 15(1963), 121-139.

472. Kobayashi, S. Hyperbolic Manifolds and Holomorphic Maps, Marcel Dekker (1970).

473. Kobayashi, S. Transformation groups in differential geometry, Ergebnisse der Mathematik und ihrer Grenzgebiete, Band 70 Springer Verlag (1972).

474. Kobayashi, S., Nagano, T. On filtered Lie algebras and geometric structures, I, J. Math. Mech. 13(1964), 875-908; II, ibid 14(1965), 513-521; III, ibid 679-706; ibid 15(1966), 163-175; V, ibid 315-328.

475. Kobayashi, S., Nomizu, K. Foundations of Differential Geometry, 2 Wiley (1969).

476. Kobayashi, S., Ochiai, I. On complex manifolds with positive tangent bundles, J. Math. Soc. Japan 22 (1970), 499-525.

477. Kodaira, K. On differential geometric method in the theory of analytic stacks, Proc. Nat. Acad. Sci. USA 39(1953), 1268-1273.

478. Kodaira, K. On Kahler varieties of restricted type( an intrinsic characterization of algebraic varieties), Ann. Math. 60(1954), 28-48.

479. Kodaira, K. Characteristic linear systems of complete continuous system, Am. J. Math. 78(1956), 716-744.

480. Kodaira, K. On the structure of compact complex analytic surfaces, I, Ann. Math. 71(1960), 111-152; II, Ann. Math. 77(1963), 563-626; III, Ann. Math. 78(1963), 1-40.

481. Kodaira, K. On the structure of compact complex surfaces, I, Amer J. Math. 86(1964), 751-798; II, ibid 88(1966), 688-721; III, ibid 90(1968), 55-83; IV, ibid 90(1968), 1048-1066.

482. Kodaira, K. A theorem of completeness of characteristic system for analytic families of compact submanifolds of complex manifolds, Ann. Math. 84(1962), 146-162.

483. Kodaira, K. On stability of compact submanifolds of complex manifolds, Am. J. Math. 85(1963), 79-94.

484. Kodaira, K. Complex structures on $S^1 \times S^3$, Proc. Nat. Acad. Sci. USA 55(1966), 240-243.

485. Kodaira, K. A certain type of irregular surface, J. Anal. Math. 19(1967), 207-215.

486. Kodaira, K. Pluricanonical systems of algebraic surfaces of general type, J. Math. Soc. Japan 20(1968), 170-192.

487. Kodaira, K. On homotopy K3 surfaces, Memoirs dedies a G. de Rham, Springer Verlag (1970), 58-69.

488. Kodaira, K. On holomorphic mappings of polydiscs into compact complex manifolds, J. Diff. Geo. 6(1971), 33-46.

489. Kodaira, K., Morrow, J. Complex Manifolds, Holt, Rinehart and Winston, (1971).

490. Kodaira, K., Nirenberg, L., Spencer, D. C. On the existence of deformation of complex analytic structures, Ann. Math. 67 (1958), 450-459.

491. Kodaira, K., Spencer, D. C. On differential geometric method in the theory of analytic stacks, Proc. Nat. Acad. Sci. USA 39(1953), 1268-1273.

492. Kodaira, K., Spencer, D. C. Division class groups on algebraic varieties, Proc. Nat. Acad. Sci. USA 39(1953), 872-877.

493. Kodaira, K., Spencer, D. C. On arithmetic genera of algebraic varieties, Proc. Nat. Acad. Sci. USA 39(1953).

494. Kodaira, K., Spencer, D. C. On the variation of almost complex structures in algebraic geometry and topology, In honour of Lefchetz, Princeton Univ. Press (1957), 139-150.

495. Kodaira, K., Spencer, D. C. A theorem of completeness for completeness for complex analytic fibre spaces, Acta Math. 100(1958), 281-294.

496. Kodaira, K., Spencer, D. C. On deformations of complex analytic structures, I, II, Ann. Math. 67(1958), 328-466; III , Ann. Math. 71(1960), 43-76.

497. Kodaira, K., Spencer, D. C. Existence of complex structures on differentiable family of deformations of compact complex manifolds, Ann. Math. 70(1959), 145-166.

498. Kodaira, K., Spencer, D. C. A theorem on completeness of characteristic systems of complete continuous systems, Amer. J. Math. 81(1959), 477-500.

499. Kohn, J. J. Harmonic integrals on strongly speudo convex manifolds, I, II, Ann. Math. 78(1963), 112-148.

500. Kohn, J. J. Sufficient conditions for subellipticity on weakly pseudo convex domains, Proc. Nat. Acad. Sci. USA 74(6) (1977), 2214-2216.

501. Kosul, J. L. Exposes sur les Espaces Homogenes Symmetriques, Publ. Soc. Math. Sao Paulo (1959).

502. Kramm, B. Uber deformationen von analytischen Abbildungskumen, Manuscripts Math. 10(1973), 163-189.

503. Ku, H. T., Mann, L. N., Sicks, J. L., Su, J. C. (Editors) Proceedings of the second conference on compact transformation groups I, II, Amherst (1971), Lecture Notes 298, 299, Springer Verlag (1971).

504. Kransov, V. A. Deformations of complex analytic modifications, Izv. Akad. Nauk SSSR Ser. Mat. 37(1973), 516-532.

505. Kuiper, N. H. A continuous function with two critical points, Bull. Am. Math. Soc., 67(1961), 281-285.

506. Kuiper, N. H. Algebraic equations for non-smoothable 8-manifolds, Publ. IHEA 33(1967), 139-155.

507. Kulkarni, R. S. On the principle of uniformisation, J. Diff. Geom. 13(1978), 109-138.

508. Kumpera, A., Spencer, D. C. Lie Equations-1: General Theory, Ann. Math. Studies, Princeton Univ. Press (1972).

509. Kuranishi, M. On. E. Cartan's prolongation theorem of exterior differential systems, Am. J. Math. 81(1957), 1-47.

510. Kuranishi, M. On the theory of continuous pseudogroups I, Nagoya Math. J. 15(1958), 225-260; II, Nagoya Math. J. 19(1961), 33-91.

511. Kuranishi, M. On a type of family of complex structures, Ann. Math. 74(1961), 262-328.

512. Kuranishi, M. On the locally complete families of complex structures, Ann. Math. 75(1962), 536-577.

513. Kuranishi, M. New proof of the existence of locally complete families of complex structures, Proc. Conf. Complex Analysis, Minneapolis (1964), Springer Verlag, (1965), 142-154.

514. Kuranishi, M. Deformations of compact complex manifolds, Univ. Montreal Press (1969).

515. Kuranishi, M. A note on families of complex structures, Global Analysis, Princeton Univ. Press and Tokyo Univ. Press (1969).

516. Kuranishi, M. Deformations of isolated singularities and $\partial_b$, Columbia Univ. Reprint (1974).

517. Kurke, G. The Castelnuovo criterion of rationality, Math. Zemetki, 11(1), (1972), 27-32.

518. Landman, A. On the Picard Lefchetz transformation for algebraic manifolds acquiring general singularities, Trans. AMS 181(1973), 89-126.

519. Landweber, P. S. Complex structures on open manifolds, Topology, 13(1974), 69-75.

520. Lang, S. Introduction to Algebraic Geometry, Interscience (1958).

521. Langmann, K. Konstructionen global Moduln und Anwendungen, Math. Z. 127(1972), 235-255.

522. Larson, M.E. On the topology of the complex projective manifolds, Invent. Math. 19(1973), 251-260.

523. Lashof, R. Equivariant smoothing theory, Bull. Am. Math. Soc. 84(1978), 1-6.

524. Lashof, R., Rothenberg, M. Microbundle and smoothing, Topology, 31(1965), 357-388.

525. Lashof, R., Rothenberg, M. Triangulations of manifolds, I, Bull. Am. Math. Soc. 75(1969). 750-754; II, ibid 755-757.

526. Laudal, O.A. Sections of functions and the problem of lifting (deforming) algebraic structures I, II, III, Oslo Univ. Preprints (1975).

527. Laufer, H.B. Normal two dimensional singularities, Ann. of Math. Studies 71, Princeton Univ. Press (1971).

528. Lawson, H.B. Jr. Codimension one foliation of sphere, Ann. Math. 94(1971), 494-503.

529. Lawson. H.B. Jr. The quantitative theory of foliations- conf board of the Math. Sci. Regional Conf. Series in Math. 27 AMS (1977).

530. Lee, R., Raymonds, R. Manifolds covering Euclidean spaces, Topology, 14(1975), 49-57.

531. Lefchetz, S. Geometrie sur les surfaces et les varietes algebriques, Mem. Sci. Math. 40(1929).

532. Lefchetz, S. Topology, Am. Math. Soc. Coll. Publ. 12(1930).

533. Lefchetz, S. Algebraic Geometry, Oxford Univ. Press (1953).

534. Lefchetz, S. Analysis situs et la geometrie algebraique, Chelsea, New York (1971).

535. Lehman, D. Sur les integrabilite des G-structures, Symposia Mathematica X (1972), 127-139.

536. Lelong, P. Fontionis plurisubharmoniques et formes differentielles positive, Gordon and Breach (1968).

537. Leray, J. L'anneau spectral et l'anneau d'homologie d'un espace localement compact et d'une application continue, J. Math. Pure Appl. 29(1950), 1-139.

538. Lewy, H. An example of a smooth linear partial differential equation without solution, Ann. Math. 66(1957), 155-158.

539. Lichtenbaum, S., Schlessinger, M. The cotangent complex of a morphism, Trans. Am. Math. Soc. 128(1967), 41-70.

540. Lichnerowicz, A. Geometrie des Groupes de Transformations, Dunod, (1958).

541. Lichnerowicz,A. Laplacien sur une variete Riemannienne et Spinors, Atti. Acad. Nat. Lincei Rend. Cl. Sci. Fis. Mat. Nat. 33(1962), 187-191.

542. Lichnerowicz,A. Varietes complexes et tensem de Bergmann, Ann. Inst. Fourier 15(1965), 345-407.

543. Lie,S. Gesammelte Abhandlungen, Band VI Oslo, Leipzig (1927), 300-365.

544. Lie,S., Engel,F. Theorie der Transformationsgruppen, 3 vols, Teubner, Leipzig, (1888-1893).

545. Liebermann,D. Pseudogroupes infinitesimaux attaches aux pseudogroupes de Lie, Bull. Soc. Math. France 87(1972), 409-425.

546. Liebermann,D., Mumford,D. Matsusaka's big theorem, Proc. Symp. Pure Math. AMS 29(1975), 513-530.

547. Liebermann,D., Serenesi,E. Semicontinuity of Kodaira dimension, Bull. Am. Math. Soc. (1975), 409-410.

548. Liebermann,D., Serenesi,E. Semicontinuity of L-dimensions, Math. Ann. 225(1977), 77-88.

549. Lipman,J. Rational singularities with applications to algebraic surfaces and unique factorisation, Publ. IHES 36 (1969), 195-280.

550. Lipman,J. Introduction to resolution of singularities in algebraic geometry, Arcata (1974), Proc. Symp. Pure Math. AMS 29(1975), 187-230.

551. Looijenga,E. A period mapping for certain semi universal deformations, Compositio Math. 30(1975), 299-316.

552. Lundel,A.T., Weingram,S. Topology of CW complexes, Van Nostrand Reinhold, New York (1969).

553. Mabuchi,T. Algebraic three folds with ample tangent bundle, Proc. Jap. Acad. 53(1977), 31-34.

554. Mabuchi,T. $C^3$ actions and algebraic threefolds with ample tangent bundle, Nagoya Math. J. 69(1978), 33-64.

555. Mabuchi,T. Almost homogeneous torus actions on varieties with ample tangent bundles, Tohoku Math. J. 30(1978), 639-651.

556. Macbeath,A.M. On a theorem of Hurewitz, Proc. Glasgow Math. Assoc. 5(1961), 90-96.

557. Macdonald,I.G. Algebraic Geometry - Introduction to Schemes, W.A. Benjamin (1968).

558. Maclane, S. Topology and logic as a source of algebra, Bull. Am. Math. Soc. 82(1976), 1-40.

559. Madsen, I. Smooth spherical space forms; Geometric Applications of Homotopy theory I, Lecture Notes 657, Springer Verlag (1975), 303-352.

560. Madsen, I., Thomas, E., Wall, C. T. C. The topological spherical space from Problem II, Topology

561. Maeda, H. Some complex structures on product of spheres, J. Fac. Sci. Univ. Tokyo, Sec. IA 21(1974), 161-165.

562. Malgrange, B. Analytic spaces, L'Enseignement Math. 14(1968), 1-28.

563. Malgrange, B. Sur l'integrabilite des structures presque complexes, Proc. Symp. Pure Math. II (1969), 289-296.

564. Malgrange, B. Equations de Lie I, J. Diff. Geom. 6(1972), 117-142.

565. Mandelbaum, R. On the topology of elliptic surfaces, Advances in Algebraic Topology, Advances in Math. Sup. Studies 5(1978).

566. Mandelbaum, R., Moishezon, B. G. On the topological structure of algebraic surfaces in $Cp^3$, Topology 15(1976), 23-40.

567. Mandelbaum, R., Moishezon, B. G. On the topology of algebraic surfaces, Topology (to appear).

568. Manin, Yu. I. Lectures on the K-functor in algebraic geometry, Russian Math. Surveys 24(1969), 1-89.

569. Markoe, A., Rossi, H. Families of strongly pseudoconvex manifolds, Lecture Notes 184, Springer Verlag (1970).

570. Maruyama, M. On classification of ruled surfaces, Lecture Notes, Kyoto Univ. (1970).

571. Maruyama, M. Moduli of stable sheaves I, J. Math. Kyoto Univ. 17(1977), II, J. Math. Kyoto Univ. 18(1978), 557-614.

572. Massey, W. Algebraic Topology, Harcourt, Brace and World (1967).

573. Massey, W. Obstructions to the existence of almost complex structures, Bull. Am. Math. Soc. 67(1961), 559-564.

574. Massey, W. On the Stiefel Whitney class of a manifold I, Am. J. Math. 82(1960), 92-102; II, Proc. Am. Math. Soc. (1962), 938-942.

575. Masuda, K. Continuity of homeomorphism of Lie algebras of vector fields, Proc. Japan Acad. 51(1975), 359-361.

576. Mather, J. Stability of $C^\infty$ mappings : I- The division theorem, Ann. Math. (1968), 89-104; II -Finitely determined map germs, Publ. IHES 35(1968), 127-156; III- Infinitesimal stability implies stability, Ann. Math. 89(1969), 254-291; IV- Classification of stable map germs by R. algebras, Publ. IHES 37(1969), 223-248; V - Transversality, Adv. Math. 4(1970), 301-331; VI - The nice dimensions, Proc. Liverpool Singularities I, Lecture Notes 112, Springer Verlag (1971), 207-255.

577. Mather, J. Loops and Foliations, Manifolds, Tokyo (1973), Tokyo. Univ. Press (1975), 175-180.

578. Matsumara, H. Commutative Algebra, Benjamin, New York (1970).

579. Matsumara, H., Oort, F. Representability of a group functors and automorphisms of algebraic schemes, Invent. Math. 4(1967), 1-25.

580. Matsusaka, T. Polarised varieties, fields of moduli and generalised Kummer varieties of polarised abelian varieties, Am. Math. 80(1958), 45-82.

581. Matsusaka, T. On canonically polarised varieties, Coll. Algebraic Geom. Tata Inst. Bombay (1969), 265-306.

582. Matsusaka, T. Theory of Q-varieties, Publ. Math. Soc. Japan, 8(1965),

583. Matsusaka, T., Mumford, D. Two fundamental theorems on deformations of polarised varieties, Am. J. Math. 86(1964), 668-684.

584. Matsushima. Y. Affine complex manifolds, Osaka J. Math. 5(1968), 215-222.

585. Matsushima, Y. On Hodge manifolds with zero first Chern class, J. Diff. Geom. 3(1969), 477-480.

586. Mattuck, A. Ruled surfaces and the Albanese mapping, Bull. Am. Math. Soc. 75(1969), 776-779.

587. Mayer, A. Families of $K_3$-surfaces, Nagoya Math. J. 48(1972), 1-17.

588. Mayer, J. Algebraic Topology, Prentice-Hall, New Jersey (1972).

589. McKean, H. P., Singer, I. M. Curvature and eigenvalues of the Laplacian, J. Diff. Geom. 1(1967), 43-69.

590. Mianishi, M., Nakamura, K. On the structure of minimal surfaces of general type with $2p_g = K^2 + 2$, Math. Kyoto Univ. 18(1978), 137-171.

591. Milnor, J. On the manifolds homeomorphic to the 7-sphere, Ann. Math. 64(1956), 394-405.

592. Milnor, J. On simply connected 4-manifolds, Symp. Int. de Topologie Algebraica, UNESCO, Mexico City, (1958), 122-128.

593. Milnor, J. On the spaces having the same homotopy type as a CW-complex, Trans. Am. Math. Soc. 90(1959), 272-280.

594. Milnor, J. The complexes which are homeomorphic but combinatorically distinct, Ann. Math. 74(1961), 575-590.

595. Milnor, J. A procedure to kill homotopy groups of differentiable manifolds, Proc. Symp. Pure Math. III(1961), 39-55.

596. Milnor, J. Microbundles, Topology, 3(1964), 53-80.

597. Milnor, J. Eigenvalues of Laplace operators on certain manifolds, Proc. Nat. Acad. Sci. USA 542(1964).

598. Milnor, J. On the Stiefel Whitney numbers of complex manifolds and of spin manifolds, Topology, 3(1965), 223-230.

599. Milnor, J. Remarks concerning spin manifolds; Differential and combinatorial topology ( in honour of Marston Morse), Princeton Univ. Press (1965).

600. Milnor, J. Singular points of complex hypersurfaces, Ann. Math. Studies 61, Princeton Univ. Press (1968).

601. Milnor, J., Husemoller, D. Symmetric bilinear forms, Ergebnisse der Mathematik und ihrer Grenzgobiete, Band 73, Springer Verlag (1973).

602. Milnor, J., Stasheff, J.D. Characteristic classes, Ann. Math. Studies 76, Princeton Univ. Press (1974).

603. Minakshisundaram, S. Eigenfunctions on Riemannian manifolds, J. Indian Math. Soc. 17(1953), 159-165.

604. Minakshisundaram, S., Pleijel, A. Some eigenfunctions of the Laplace operator on Riemannian manifolds, Canad. J. Math. 1(1949), 242-256.

605. Miyanishi, M. Unirational quasi-elliptic surfaces, Jap. J. Math. 3(1977), 395-416.

606. Miyaoka, Y. Kahler matrices on elliptic surfaces, Proc. Jap. Acad. Sci. 50(1974), 533-536.

607. Miyaoka, Y. On Chern numbers of surfaces of general type, Invent. Math. 42(1977), 225-237.

608. Moise, E.E. Geometric Topology in Dimensions 2 and 3, A Springer Graduate Text No. 47(1977).

609. Moishezon, B. G. Resolution theorems for compact complex spaces with sufficiently large field of meromorphic functions, Math. USSR Izv. 1(1967), 1331-1356.

610. Moishezon, B. G. A criterion for projectivity of complete algebraic abstract varieties, AMS Translations 63(1967), 1-50.

611. Moishezon, B. G. On n-dimensional compact varieties with n algebraically independent meromorphic functions I, II, III, AMS Translations Ser. 2 No. 63(1967), 51-177.

612. Moishezon, B. G. In Proc. Int. Congr. Math. Nice (1970).

613. Moishezon, B. G. Singular Kahlerian spaces, Manifolds, Tokyo (1973), Hattori, A (Editor), Univ. Tokyo Press (1975), 343-353.

614. Moishezon, B. G. Some estimates in the topology of simply connected algebraic surfaces, Bull. Am. Math. Soc. 83(1977), 741-744.

615. Moishezon, B. G. Complex Surfaces, Lecture Notes 603, Springer Verlag (1978).

616. Moise, E. E. Affine structures on 3 manifolds V: The triangulation theorem and Hauptvernutung, Ann. Math. 56(1952), 96-114.

617. Molino, P. Theorie des G-Structures: Lie Probleme d'equivalence, Lecture Notes 588, Springer Verlag (1978).

618. Montgomery, D., Zippin, L. Transformation Groups, Interscience Tracts No. 1 (1955).

619. Moolgavkar, S. H. The existence of a universal germ of deformations for elliptic pseudogroup structures on compact manifolds. Trans. Am. Math. Soc.

620. Mori, S. Projective manifolds with ample tangent bundles, Reprint, Harvard Univ. (1979).

621. Morita, S. Smoothability of PL manifolds is not topological invariant, Manifolds, Tokyo(1973), 51-56.

622. Morita, S. Almost complex manifolds and Hirzebruch invariants for isolated singularities in complex spaces, Math. Ann. 211(1974).

623. Morita, S. Topological classification of complex structures on $S^1 \times \Sigma^{2n-1}$, Topology, 14(1975), 13-22.

624. Morrey, C. B. Jr. The analytic embedding of abstract real analytic manifolds, Ann. Math. 68(1958), 159-201.

625. Morrow, J. Survey on some results on Kahler manifolds, Global Analysis, Princeton Univ. Press and Tokyo Univ. Press (1969).

626. Morrow, J. Minimal compactification of $C^2$, Rice Univ. Studies 59, No. 1(1973), 97-112.

627. Morrow, J., Rossi, H. Some theorems of algebraicity for complex spaces, J. Math. Soc. Japan 27(1975), 167-183; Connections, J. Math. Soc. Japan 29(1977), 783.

628. Morse, M., Cairns, S. Critical Point Theory in Global Analysis and Differential Topology, Academic Press (1969).

629. Mumford, D. The topology of normal singularities of an algebraic surface and criterion for simplicity, Publ. Math. IHES, Bures-sur Yvette 89(1961), 5-22.

630. Mumford, D. Pathologies of modular algebraic geometry, Amer. J. Math. 83(1961), 339-342; Further pathologies in algebraic geometry, Amer. J. Math. 84(1962), 642-648, Pathologies III, Amer. J. Math. 89(1967), 94-104.

631. Mumford, D. Projective invariants of projective structures and applications, Proc. Int. Congr. Math. Stockholm (1962), 526-530.

632. Mumford, D. The canonical ring of an algebraic surface, Ann. Math. 76(1962), 612-615.

633. Mumford, D. Geometric Invariant Theory, Springer Verlag (1965).

634. Mumford, D. Picard groups of moduli problems, Arithmetical Algebraic Geometry, Harper and Row, New York (1965), 33-81.

635. Mumford, D. Lectures on curves on an algebraic surface, Ann. Math. Studies 59, Princeton Univ. Press (1966).

636. Mumford, D. Enriques Classification of Surfaces in Characteristic p: Global Analysis (in honour of Kodaira), Princeton Univ. Press (1969).

637. Mumford, D. Introduction to the theory of moduli, Algebraic Geometry, Oslo, (1970).

638. Mumford, D. Abelian Varieties, Tata Inst. and Oxford Univ. Press (1970).

639. Mumford, D. An analytic construction of degenerating abelian varieties over complete rings, Comp. Math. 24(1972), 239-272.

640. Mumford, D. Curves and their Jacobians, Univ. Mich. Press, Ann Arbor (1975).

641. Mumford, D. Algebraic Geometry I, Complex Projective Varieties, Grundlehren der mathematischen, Wissenschaften No. 221, Springer Verlag (1976).

642. Mumford, D. Stability of projective varieties: L'ensignment Mathematique XXIII (1977), 39-110.

643. Mumford, D. Some footnotes to the work of C. P. Ramanujan, In C. P. Ramanujan: A Tribute, Tata Institute (1978), 247-262.

644. Mumford, D. Lectures on surfaces given at the Ramanujan Institute, January 1979.

645. Mumford, D., Newstead, P. E. Periods of a moduli space of bundles on curves, Amer. J. Math. 90(1968), 1200-1208.

646. Mumford, D., Oort, F. Deformations and liftings of finite commutative group schemes, Invent. Math. 5(1968), 317-334.

647. Mumford, D., Suominen, K. Introduction to the theory of moduli algebraic geometry, Oslo(1970), Wolters-Noordhoff (1972), 171-222.

648. Munkers, J. Obstructions to the smoothing of piecewise linear homeomorphisms, Ann. Math. (1960), 521-554.

649. Munkers, J. Obstructions to imposing differentiable structures, Illinois J. Math. 18(1964), 361-376.

650. Munkers, J. Higher obstructions to smoothing, Topology, 4(1965), 27-45.

651. Munkers, J. Concordance is equivalent to smoothability, Topology, 5(1966), 371-389.

652. Munkers, J. Concordance of differentiable structures, Mich. Math, J. 14(1967), 183-191.

653. Murre, J. P. On contravariant functors from the category of presheaves into the category of abelian groups, Publ. IHES 23(1964).

654. Murre, J. P. Reduction of the proof of the nonrationality of a non-singular cubic threefold to a result of Mumford, Comp. Math. 27(1973), 63-82.

655. Murre, J. P. Some results on cubic threefolds, Lecture Notes 412, Springer Verlag (1974), 140-164.

656. Nagata, M. A general theory of algebraic geometry over Dedekind domain, Am. J. Math. 78(1956), 78-116.

657. Nagata, M. On the imbedding of abstract surfaces in projective variety, Mem. Coll. Sci. Kyoto 30(1957), 231-235.

658. Nagata, M. On rational surfaces I, Coll. Sci. Univ. Kyoto 32(1960), 351-370; II, Coll. Sci. Univ. Kyoto 33(1961), 271-293.

659. Nagata, M. Imbedding of an abstract variety in a complete variety, J. Math. Kyoto Univ. 2(1962), 1-10.

660. Nagel, A. Flatness criterion for modules of holomorphic functions over $\theta_n$, Duke Math. J. 40(1973), 433-448.

661. Nakai, Y. A criterion of an ample sheaf on a projective schems, Am. J. Math. 85(1963), 14-21.

662. Nakamura, I. Complex parallelizable manifolds and their small deformations, J. Diff. Geom. 10(1975), 85-112.

663. Nakamura, I. On moduli of stable quasiabelian varieties, Nagoya Math. J. 58(1975), 149-214.

664. Nakamura, I., Ueno, K. An addition formula for Kodaira dimensions of analytic fibre bundles whose fibres are Moishezon manifolds, J. Math. Soc. Jap. 25(1973), 363-371.

665. Nakano, S. On complex analytic vector bundles, J. Math. Soc. Jap. 7(1955), 1-12.

666. Nakano, S. On the inverse monoidal transformation, Publ. Res. Inst. Kyoto Univ. 6(1971), 483-502.

667. Namba, M. On maximal families of compact complex submanifolds of a complex manifold, Tohoku Math. J. 24(1974), 581-609.

668. Namba, M. On deformations of automorphisms of groups of compact complex manifolds, Tohoku Math. J. 26(1974), 273-283.

669. Namba, M. Moduli of open holomorphic maps of compact complex manifolds, Math. Ann. 220(1976), 65-76.

670. Namba, M. On families of effective divisors on algebraic manifolds Proc. Japan Acad. 53A(1977), 206-209.

671. Namikawa, Y. Studies of degenerations, Lecture Notes 412, Springer Verlag (1974), 165-210.

672. Namikawa, Y. A new compactification of the Sigel space and degeneration of abelian varieties I, Math. Ann. 221(1976) 97-141; II ibid 201-241.

673. Namikawa, Y., Ueno, K. The complete classification of fibre pencils of curves of genus 2, Manuscripta Math. 9(1973), 163-186.

674. Narasimhan, M. S. Variations of complex structures on an open Riemann surface, Ann. Inst. Fourier, Grenoble, Tome XI(1961), 493-514.

675. Narasimhan, M. S. Vector bundles over Riemann surfaces, Complex Analysis and its Applications, Vol. III, IAEA(1977).

676. Narasimhan, M. S., Ramadas, T. R. Geometry of SU(2) gauge fields, Commum. Math. Phys. 67(1979), 121-136.

677. Narasimhan, M. S., Ramanan, S. Universal connections I, Am. J. Math. 83(1961), 536-572; II, Am. J. Math. 85(1963), 223-231.

678. Narasimhan, M. S., Ramanan, S. 1) Moduli of vector bundles on a compact Riemann surface, Ann. of Math. 89(1969), 19-51.
2) Deformations of moduli spaces of vector bundles on an algebraic curve, Ann. of Math. 107(1975), 391-417.

679. Narasimhan, M. S., Seshadri, C. S. Stable and unitary vector bundles on a compact Riemann surface, Ann. Math. 82(1965), 540-567.

680. Narasimhan, M. S., Simha, R. Manifolds with ample canonical class, Invent. Math. 5(1968), 120-126.

681. Narasimhan, R. Embedding of holomorphically complete complex spaces, Am. J. Math. 82(1960), 917-934.

682. Narasimhan, R. The Levi problem for complex spaces I, Math. Ann. 142(1961), 355-365; II, Math. Ann. 146(1962), 195-216; Pric. Inst. Congr. Math. Stockholm (1962), 385-388.

683. Narasimhan, R. Introduction to the theory of analytic spaces, Lecture Notes 25, Springer Verlag (1966).

684. Narasimhan, R. Analysis on Real and Complex Manifolds, North-Holland, Amsterdam (1968).

685. Narasimhan, R. Compact analytical varieties, L'enseignment Mathematique XIV(1968), 75-98.

686. Narasimhan, R. Grauert's Theorem on Direct Images of Coherent Sheaves, University of Montreal Press (1968).

687. Narasimhan, R. Several Complex Variables, The University of Chicago Press (1971).

688. Narasimhan, R. The Levi problem and pseudoconvex domains, A Survey, L'enseignment Mathematique, XXIV(1978), 161-172.

689. Nash, J. Real algebraic manifolds, Ann. of Math. 56(1952), 405 -421.

690. Nash, J. The imbedding problem for Riemannian manifolds, J. Indian Math. Soc. 17(1953), 159-165.

691. Newlander, A, Nirenberg, L. Complex analytic coordinates in almost complex manifolds, Ann. Math. 65(1957), 391-404.

692. Newman, M. A. H. Topology of plane sets, Camb. Univ. Press (1935).

693. Newman, M. A. H. The engulfing for topological manifolds, Ann. Math. 84(1966), 555-571.

694. Newstead, P. E. Rationality of moduli spaces of stable bundles, Ann. Math. 215(1975), 251-268.

695. Newstead, P. E. Introduction to Moduli Problem and Orbit Spaces, Tata Inst. Lecture Notes, Bombay (1978).

696. Nijenhuis, A. Vector form methods and deformations of complex structure, Diff. Geometry, Proc. Symp. Pure Math. III, AMS (1961), 87-93.

697. Nijenhuis, A., Richardson, R. Cohomology and deformations in graded Lie algebras, Bull. Am. Math. Soc. 72(1966), 1-29.

698. Nijenhuis, A., Woolf, J. Some integration problems in almost complete manifolds, Ann. Math. 77(1963), 424-483.

699. Nirenberg, L. 1) Partial differential equations with applications to geometry, Lectures on Modern Mathematics, 2, Wiley (1964).
2) Lectures on partial differential equations, CBMS Series AMS (1977).

700. Nishino, T. Sur les families de surfaces analytiques, J. Math. Kyoto Univ. 1(1962), 357-377.

701. Nori, M. Varieties with no smooth embeddings, In C. P. Ramanujan - A Tribute: Publ. of the Tata Institute, Bombay (1978).

702. Norton, A. Non-separation in the moduli of complex vector bundles, Math. Ann. 235(1978), 1-16.

703. Novikov, S. P. Homotopy equivalent smooth manifolds, Am. Math. Soc. Transl. 46(1965), 271-396.

704. Novikov, S. P. Topological invariance of rational Pontrjagin classes, Dokl. Akad. Nauk SSSR 163(1965), 298-300.

705. Novikov, S. P. On manifolds with free abelian fundamental groups and applications (Pontrjagin classes, smoothings and highdimensional knots), Izv. Akad. Nauk SSSR 30(1966), 208-246.

706. Novikov, S. P., Pjateckii Sapiro, I. I., Safarevich, I. R. Fundamental directions in the development of algebraic topology and algebraic geometry, Uspehi Mat. Nauk 19(1964), No. 6(120), 75-82.

707. Ochiai, T. On compact Kahler manifolds with positive holomorphic bisectional curvature, Proc. Symp Pure Math. 27(1975), 113-123.

708. Oda, T. Lectures on Torus Embedding and Applications,

Tata Inst. Lecture Notes, Bombay (1978).

709. Oka, K. Sur les fonctions analytique de plusieur, Iwanami Shoten (1961).

710. Okano, T. A remark on complex analytic families of complex tori, Pub. Res. Inst. Math. Sci. Ser. A 3(1967/68), 289-297.

711. Omoto, H., Nakano, S. Local deformations of isolated singularities associated with negative line bundles over abelian varieties, Nagoya Math. J. 75(1979), 41-70.

712. Oort, F. Fine and Coarse Moduli Schemes are Different, Springer Lecture Notes 586 (1977).

713. Oort, F., Tate, J. Group schemes of prime order, Ann. Ec. Norm. Sup. 2(1969), 1-21.

714. Oort, F. Finite group scheme, local moduli for abelian varieties and lifting problems, Proc. 5th Nordic Summer School in Math. Oslo (1970), 223-254.

715. Oort, F. (Editor) Algebraic Geometry, Oslo(1970), Wolters-Noordhoff (1972).

716. Oort, F. Singularities of moduli schemes for curves of genus three, Proc. Kon. Ned. Akad. 78(1975), 170-174.

717. Oort, F. Singularities of coarse moduli schemes: Seminaire d'Algebre Paul Dubreil 1975-76, Lecture Notes 586, Springer Verlag (1977), 61-76.

718. Ozaki, S., Oda, T. On some properties of flat families, Sci. Rep. Tokyo, Kyoiku Daigaku A 10(1970), 297-300.

719. Palamodov, V. P. Moduli in versal deformations of complex spaces, Springer Lecture Notes 683(1978), 74-115.

720. Palamodov, V. P. On the existence of versal deformations of complex spaces, Sov. Math. Dokl. 13(1972), 1246-1250.

721. Palamodov, V. P. Deformations of complex spaces, Russian Math. Surveys, 31(3), (1976), 129-197.

722. Palais, R. (Editor) Seminar on Atiyah-Singer Index Theorems, Ann. Math. Studies 57, Princeton Univ. Press (1965).

723. Patodi, V. K. Curvature and the fundamental solutions of the heat equations, J. Ind. Math. Soc. 34(1970), 269-285.

724. Patodi, V. K. An analytic proof of the Riemann-Roch-Hirzebruch theorems for Kahler manifolds, J. Diff. Geom. (1971), 251-283.

725. Patodi, V.K. Curvature and the eigenforms of the Laplace operator, J.Diff.Geom.5(1971),233-249.

726. Persson, U. On degeneration of algebraic surfaces, Memoirs AMS 189(1977).

727. Peters, C.A.M. On two types of surfaces of general type with vanishing genus, Invent.Math.32(1976),33-47.

728. Peters, C.A.M. On certain examples of surfaces with $p_g=0$ due to Burniat, Nagoya Math.J.66(1977),109-119.

729. Peters, C.A.M. Holomorphic automorphisms of compact Kahler surfaces and their induced actions in cohomology, Invent. Math. 52(1979),143-148.

730. Pflug, R.P. Holomorphic functions of polynomial growth and applications: Hyperbolic Complex Analysis, Pub.Ramanujan Inst.No.4(1979),141-163.

731. Pham, F. Formules de Picard-Lefchetz generalise et ramification des integrals, Bull.Soc.Math.France, 93(1965),333-367.

732. Phillips, A. Foliations on open manifolds I, Comm.Math.Helv. 43(1968),204-211; II, Comm.Math.Helv.44(1969), 367-370.

733. Pittie, H. Complex and almost complex four-manifolds, Complex Analysis and its Applications, Vol.3 IAEA (1977).

734. Pittie, H. Characteristic classes of foliations, Research Notes in Mathematics. Pitman Publishing, London (1978).

735. Pjatecki, Sapiro, I.I., Safarevich, I.R. A Torelli theorem for algebraic surfaces of type K3, Math.USSR-Izv.5(1971),547-588.

736. Poenaru, V. On handle bodies, Topology, 9(1970),169-181.

737. Poincare, H. Analysis Situs, J.Ecole Polytech.Paris 21(1890),1-12.

738. Poljakov, P.L. The triviality of families of complex spaces, Math. Sb.(N.S.),79(121)(1969),357-367.

739. Pollack, A. The integrability problem for G-structures, J.Diff. Geom.9(1974),268-279.

740. Pollack, A. The integrability problem for pseudogroup structures, J.Diff.Geom.9(1974),355-390.

741. Pommaret, J.F. Physics and deformation theory of Lie algebras, Lecture Notes in Physics 50, Springer Verlag (1976), 523-536.

742. Ponomarev, D.A. Deformations of principal analytic bundles over spaces with solvable singularities, Mat. Sb. (N.S.)88(130) (1972), 623-629.

743. Pontrjagin, L.S. On the classification of 4-dimensional manifolds, Uspehi Mat. Nauk 4(1949), 4(32), 157-158.

744. Pontrjagin, L.S. Some topological invariants of closed Riemann manifolds, AMS Transl. 49(1951).

745. Popp, H. The singularity of moduli scheme of curves, J. Number Theory 1(1969), 90-107.

746. Popp, H. Fundamentalgruppen algebraischen Mannigfaltigkeiten, Lecture Notes (1970), Springer Verlag (1970).

747. Popp, H. On moduli of algebraic varieties I, Invent. Math. 22(1973), 1-40; II, Compositio Math. 28(1974), 51-81; III ibid 31(1975), 237-258.

748. Popp, H. Modulraume algebraischer Mannigfaltigheiten, Lecture Notes 412, Springer Verlag (1974), 219-242.

749. Popp, H. Moduli theory and classification theory of algebraic variety, Lecture Notes 620, Springer Verlag (1977).

750. Popp, H. Universalle Familien Kurven, Abh. Math. Seminar Universitat Hamburg 47(1978), 228-235.

751. Postnikov, M.M. Investigations in homotoph theory of continuous maps I, AMS Transl. Series 2 Vol, 7(1957), 1-34; II, AMS Transl. Ser. 2 Vol. 9(1959), 115-153.

752. Prestel, A. Die elliptischen Fixpunkte der Hilbertschen Modulgruppen, Math. Ann. 117(1968), 181-209.

753. Pourcin, J. Theorem de Douady au-dessus de S. Ann. Scuola Sup. Pisa 23, 3(1969).

754. Pourcin, J. Deformation de singularities isoles, Astrique No. 16, Soc. Math. France (1974), 161-173.

755. Price, J.M. Towards classifying all manifolds, Mathematical Chronicle, Vol. 7(1978), 1-47.

756. Que Ngo Van Non-abelian Spencer cohomology and deformation theory, J. Diff. Geom. 3(1969), 165-211.

757. Quillen, D. Formal properties of over determined systems of linear partial differential equations, Theses, Harvard (1964).

758. Quillen, D. On the (Co)homology of commutative rings, Proc. Symp. Pure Math. 17(1970), 65-87.

759. Raghunathan, M. S. Deformations of linear connections and Riemannian manifolds, J. Math. Mech. 13(1964), 97-123; Addendum 1043-1046.

760. Ramanan, S. The moduli spaces of vector bundles over an algebraic curve, Math. Ann. 200(1973), 69-84.

761. Ramanathan, A. Stable principal bundles on a compact Riemann surface, Math. Ann. 213(1975), 129-132.

762. Ramanathan, K. G. (Ed) C. P. Ramanujam- A Tribute, Tata Inst. Publ. Bombay (1978).

763. Ramanujam, C. P. A note on automorphism groups of algebraic varieties, Math. Ann. 156(1964), 25-33.

764. Ramanujam, C. P. On a certain purity theorem, J. Indian Math. Soc. 34(1970), 1-10.

765. Ramanujam, C. P. A topological characterisation of the affine plane as an algebraic variety, Annals of Math. 94(1971), 69-88.

766. Ramanujam, C. P. Remarks on the Kodaira vanishing theorem, J. Indian Math. Soc. 36(1972), 41-51.

767. Ramanujam, C. P. Supplement to remarks on the Kodaira vanishing theorem, J. Indian Math. Soc. 38(1974), 121-124.

768. Ramis, J. P., Ruget, G. Complexes dualisants et theorem de dualite en geometrie analytique complexe, Pub. Math. IHES 38(1970), 77-91.

769. Ramis, J. P., Ruget, G., Verdier, J. L. Dualite relative en geometrie analytique complexe, Invent. Math. 13(1971), 281-283.

770. Rauch, H. E. Weierstrass points, branch points and the moduli of Riemann surfaces. Comm. Pure Appl. Math. 12(1959), 543-560.

771. Rauch, H. E. The singularity of modulus space, Bull. Am. Math. Soc. 68(1962), 390-394.

772. Rauch, H. E. A transcendental view of the space of algebraic Riemann surfaces, Bull. Am. Math. Soc. 71(1965), 1-39.

773. Rauch, H. E. Variational methods in the problem of moduli of Riemann surfaces, In Contributions to Function Theory Tata Institute, Bombay (1960), 17-40.

774. Raymond, F., Wigner, D. Construction of aspherical manifolds, Geometric Applications of Homotopy Theory I, Lecture Notes 657, Springer Verlag (1975).

775. Raynaud, M. Families de fibres vectoriels sur une surface de Riemann, Sem. Bourbaki, 1966/67, Expose 316(1968).

776. Raynaud, M. Flat modules in algebraic geometry, Proc. Algebraic Geometry Oslo, (1970), 255-275.

777. Raynaud, M. Contre-example an vanishing theorem en Characteristique $p > 0$, in C. P. Ramanujam, A Tribute, Tata Institute, Bombay (1978), 273-278.

778. Raynaud, M. Specialisation der fontern de Picard, Publ. IHES 38(1974), 27

779. Rees, G. Sur certaines propertes topologique des varietes deuilletees, Actualities Sci. 1183, Herman Paris (1952).

780. Reid, M. Bogomolav's theorem $C_1^2 \leq 4C_2$, Proc. Int. Collo. Algebraic Geometry, Kyōto (1978).

781. Reid, M. Surfaces with $p = 0$ and $k^2 = 1$, J. Fac. Sci. Math. Tokyo Univ. 25(1978), 75-92.

782. Remmert, R. Holomorphe and meromorphe Abbildungen Komolexer Raume, Math. Ann. 133(1975), 328-360.

783. Remmert, R. Habilitationsschrift, Munster (1956).

784. Rham, G. De. Varieties differentiables, Acta. Sci. et Ind. 122 Hermann, Paris 1954.

785. Rham, G. De., Kodaira, K. Harmonic Integrals, Institute for Advanced Study, Princeton.

786. Riemann, B. Theorie der Ahelschen Funktionen, J. Reine Angew Math. 54(1857), 115-155. See Collected Works, Dover Reprint, New York.

787. Riemannschneider, O. Uber die Anwendung algebraicher Methoden in Deformation theorie Complexer Raume, Math. Ann. 187(1970), 40-55.

788. Riemannschneider, O. Characterizing Moishezon spaces by almost positive coherent analytic sheaves, Math. Z. 123(1971), 263-284.

789. Riemannschneider, O. Deformations of rational singularities and their resolutions, Rice Univ. Studies 59, Part I, (1973), 119-130.

790. Riemannschneider, O. A generalisation of Kodaira embedding theorem, Math. Ann. 200(1973), 99-102.

791. Rim, D. S. Torsion differentials and deformation, Trans. Am. Math. Soc. 169(1972), 267-278.

792. Rohlin, V. A. A new result in the theory of four-dimensional manifolds, Doklady Akad. Nauk SSSR 84(1952), 221-224.

793. Rohlin, V.A. New examples of four dimensional manifolds, Doklady Akad Naum SSR 162(1965), 273-276.

794. Rohrl, H. Holomorphic fibre bundles over Riemann surfaces, Bull. Am. Math. Soc. 68(1962), 125-160.

795. Roquette, P. Abstratzung der Automorphismenazahal von Funktionen Korpen bei Primzahl Charakteristik, Math. Zeit. 117(1970), 157-169.

796. Roth, L. Algebraic threefolds, Springer Verlag (1955).

797. Rourke, C. P., Sanderson, B. J. Introduction to Piecewise Linear Topology, Eregebnisse der Mathematik und ihrer Grenzgebiete, Band 69, Springer (1972).

798. Roydan, H. L. Automorphisms and Isometries of Teichmuller Spaces, Ann. Math. Studies 66(1971), 369-383.

799. Roydan, H. L. Remarks on Kobayashi metric, Proc. Symp. Pure Math. AMS 27(1975).

800. Rudikov, Safarevich, I. R. Inseparable morphisms of algebraic surfaces, Maht. USSR Izv. 10(1976), 1205

801. Ruget, G. Le Probleme des modules pourles fibres vectorels holomorphes sur un espace analytique complexe compact donne I : Le cas d'une base lisse quelques Problemes de modules, Astriques 16. Soc. Math. France (1974), 250-254.

802. Safarevich, I. R., et. al. Algebraic surfaces, Proc. Steklov Inst. AMS Transl. (1967).

803. Safarevich, I. R. Basic Algebraic Geometry, Springer Verlag (1974).

804. Saint Donat, B. Projective models of $K_3$ surfaces, Am. J. Math. 96(1974), 602-639.

805. Saito, K. Einfach elliptische singulariaten, Invent. Math. 23(1974), 289-325.

806. Sakai, F. Kodaira dimension of complements of divisors, Complex Analysis and Algebraic Geometry ( in honour of Kodaira), Cambridge Univ. Press (1977), 239-258.

807. Sakai, T. On theeigenvalues of the Laplacian and the curvature of Riemannian manifolds, Tohoku Math. J. 23(1971), 589-603.

808. Sakane, Y. On compact complex affine manifolds, J. Math. Soc. Japan 29(1977), 135-149.

809. Samelson, H. Topology of Lie Groups, Bull. AMS 58(1952), 2-37.

810. Samuel, P., Zariski, O. Commutative Algebra, Vol. I and II, Van Nostrand, (1960).

811. Satake, I. Holomorphic imbeddings of symmetric domains into a Siegel space, Amer. J. Math. 87(1965), 425-461.

812. Sato, H. Constructing manifolds by homotopy equivalences I, Ann. Inst. Fourier 22(1972), 271-287; II, Manifolds, Tokyo (1973), Tokyo Univ. Press (1975), 77-83.

813. Scharlemann, M.G., Siebermann, L.G. The Hauptrermatung for smooth singular homeomorphisms, Manifolds, Tokyo (1973), Math. Soc. Jap. and Princeton Univ. Press (1975), 85-91.

814. Schlessinger, M. Infinitesimal deformations of singularities, Ph.D. Thesis, Harvard Univ. (1964).

815. Schlessinger, M. Functors of Artin Rings, AMS Transl. 130(1968), 208-222.

816. Schlessinger, M. On rigid singularities, Rice. Univ. Studies 59(1973), 129-146.

817. Schmid, W. Variation of Hodge structures: the singularities of the period mapping, Invent. Math. 22(1973), 213-319.

818. Schulman, H. Characteristic classes and foliations, Ph.D. Thesis, Univ. Calif. Berkeley, California, USA (1972).

819. Schuster, H.W. Deformationen analytischen Algebren, Invent. Math. 6(1968), 262-274.

820. Schuster, H.W. Zur Theorie der Deformationen Kompakter Komplexer Raume, Invent. Math. 9(1969), 284-294.

821. Schuster, H.W. Uber die Starheit Kompakter Komplexer Raume, Manuscripta Math. 1(1969), 129-137.

822. Schwarz, L. Complex Manifolds, Tata Inst. Lecture Notes, Bombay (1950).

823. Schwarzenberger, L.E. Topics in Differential Topology, Publ. Ramanujan Inst. Univ. Madras (1972).

824. Schweitzer, P.A. (Ed) Differential Topology, Foliations and Gel'fand Fuks Cohomology, Rio de Janeiro (1976), Lecture Notes 652, Springer Verlag (1978).

825. Scott, G.P. On sufficiently large 3-manifolds, J. Math. Oxford Series (2) 23(1972), 159-172; Correction-ibid (2) (1973) 527-529.

826. Seeley, R.T. Complex powers of an elliptic operator, singular integral, operator, singular integrals, Proc. Symp. Pure Math. 10 AMS (1967), 288-307.

827. Serenesi, E. L-dimension and deformation, Proc. Symp. Pure Math. AMS 30(1977), 285-288.

828. Serre, J. P. Quelques Problems globeaux relatifs on x varietes de Stein, Colloq. Funct. de plus. variables, Brussels.

829. Serre, J. P. Un theoreme de dualite, Comm. Math. Helv. 29(1955), 9-26.

830. Serre, J. P. Faisceaux algebriques coherents (popularly referred to as FAC), Ann. Math. 61(1955), 197-278.

831. Serre, J. P. Geometrie algebraique et geometrie analytique (polularly known as GAGA), Ann. Inst. Fourier 6(1956), 1-42.

832. Serre, J. P. Sur les cohomologie des varietes algebri ques, J. de Math. Pure et Appl. 36(1957), 1-16.

833. Serre, J. P. Sur la topologie des varietes algebriques en characteriste p, Symp. Int. de Topologia Algebraica, Mexico (1958), 24-53.

834. Serre, J. P. Quelques properties des varietes abelinnes encharacteristique p, Am. J. Math. 80(1959), 715-739.

835. Serre, J. P. Formes Bilinear symmetric Entiers, Seminar Henri Cartan, 1961/62, No. 14.

836. Serre, J. P. Examples de varietes projectives en car p nou relevables en car 0, Proc. Nat. Acad. Sci. USA 47(1961), 108-109.

837. Serre, J. P. Example de variete projective conjugate non-homeomorphes, C. R. Acad. Sci. Paris 258(1964), 4194-4196.

838. Seshadri, C. S. Space of unitary vector bundle on a compact Riemann surface, Ann. Math. 85(1967), 303-336.

839. Seshadri, C. S. Moduli of $\pi$- vector bundles over an algebraic curve, Questions on Algebraic variety, CIME, Rome (1970).

840. Seshadri, C. S. Quotient space modulo reductive algebraic groups, Ann. Math. 95(1972), 511-556.

841. Seshadri, C. S. Theory of Moduli, Proc. Conf. Algebraic Geometry, Arcata, Calif. Am. Math. Soc. 29(1975), 263-304.

842. Seshadri, C. S. Desingularisation of moduli varieties of vector bundles on curves, Proc. Int. Conf. Algebraic Geometry Kyoto(1977).

843. Seshadri, C. S. Geometric reductivity over arbitrary base, Advances in Math. 26(1977), 225-274.

844. Severi, F. Uber die Grundlagen der algebraischen geometrie, Hamb. Abh. 9(1933), 335-364.

845. Shanks, M. E., Pursell, L. E. The Lie algebra of a smooth manifold, Proc. Am. Math. Soc. 5(1954), 468-472.

846. Shimura, G. Moduli and fibre systems of abelian varieties, Ann. Math. 83(1966), 294-338.

847. Shimura, G. Algebraic varieties without deformation and the Chow variety, J. Math. Jap. 20(1968), 336-341.

848. Shioda, T. Elliptic modular surfaces I, Proc. Jap. Acad. 45(1969), 786-790; II, ibid 833-837.)

849. Shioda, T. The period map for abelian surfaces, J. Fac. Sci. Math. Univ. Tokyo 25(1978), 47-59.

850. Shioda, T., Inose, H. On singular $K_3$ surfaces, Complex Analysis and Algebraic Geometry (in honour of Kodaira), Iwanami Shoten and Camb. Univ. Press (1978).

851. Siegel, C. L. Analytic Functions of Several Complex Variables, Inst. Advanced Study, Princeton (1948).

852. Siegel, C. L. Meromorphic Functionen auf Kompakten analytischen Mannigfaltigkeiten, Nach. Akad. Wiss. Gottingen (1965), 71-77.

853. Siegel, C. L. Topics in Complex Function Theory in 3 volumes, Wiley-Interscience, New York (1969-73).

854. Simha, R. R. Uber die Kritschen werte gewisser holomorpher Abbildungen, Manus. Math. 3(1)(1970), 97-104.

855. Simha, R. R. Algebraic varieties biholomorphic to $C^* \times C^*$, Tohoku Math. Jour. 30(1978), 455-461.

856. Singer, I. M. Some remarks on the Gribov ambiguity, Comm. Math, Physics 60(1978), 7-12.

857. Singer, I. M., Sternberg, S. The infinite groups of Lie and Cartan, J. d'Analyse Math. 15(1965), 1-114.

858. Siu, Y. T. The 1-convex generalization of Grauert's direct image theorem, Math. Ann. 193(1971), 203-214.

859. Siu, Y. T. Pseudoconvexity and the problem of Levi, Bull. Amer. Math. Soc. 84(1978), 481-512.

860. Siu, Y. T., Yau, S. T. Complex Kahler manifolds with non positive curvature of faster than quadratic decay, Ann. Math. 105(1975), 225-264; Errata, Ann. Math. 109(179), 621-623.

861. Smale, S. Generalised Poincare conjecture in dimension greater than four, Ann. Math. 74(1961), 391-406.

862. Smale, S. Differential and Combinatorial Structures on Manifolds, Ann. of Math. 74(1961), 498-502.

863. Smale, S. On the structure of manifolds, Amer. Jour. Math. 84(1962), 387-399.

864. Spencer, D. C. Some remarks on homological analysis and structures, Differential Geometry Symposium in Pure Math. Am. Math. Soc. IV (1961).

865. Spencer, D. C. Deformations of structures on manifolds defined by transitive continuous pseudo groups I and II, Ann. Math. 76(1962), 306-455.

866. Spivak, M. Space satisfying Poincare duality, Topology, 6(1967), 77-101.

867. Srinivasacharyulu, K. Topology de certaines varietespresque Compleses et compeacs, Rev. Roumaine Math. Pures appl. 15(1970), 1535-1540.

868. Stallings, J. Group theory and 3-manifolds, Yale Univ. Monograph No. 4, Yale Univ. Press (1971).

869. Stanton, K. Nancy, Deformation of complex manifolds with fixed boundary structure, Math. Ann. 230(1977), 227-243.

870. Stasheff, J. D. Continuous cohomology groups and classifying spaces, Bull. Am. Math. Soc. 84(1978), 513-530.

871. Steenrod, N. Cohomology Operations, Ann. Math. Studies 50, Princeton Univ. Press (1962).

872. Steenrod, N. The Topology of Fibre Bundles, Princeton Univ. Press (1951).

873. Stein, K. Meromorphic mappings, L8Enseignment Math. II 14(1968), 29-46.

874. Sternberg, S. Lectures on Differential Geometry, Prentice Hall New Jersey (1964).

875. Sullivan, D. Triangulation homotopy equivalence, Ph. D. Thesis, Princeton (1965).

876. Sullivan, D. Geometric Topology, Seminar Notes, Princeton (1969).

877. Sullivan, D. Differential forms and topology of manifolds, Manifolds, Tokyo(1973); Tokyo Univ. Press (1975), 37-49.

878. Sundararaman, D. Normal filterations of $H(M, \Theta)$, J. Diff. Geom. 8(1973), 225-248.

879. Sundararaman, D. Normal families of complex structure, J. Indian Math. Soc. 39(1975), 149-154.

880. Sundararaman, D. Deformations and classification of compact complex manifolds, Complex Analysis and its Applications, Vol. III, IAEA (1977), 133-180.

881. Sundararaman, D. On the Kuranishi space of holomorphic principle bundle over a compact complex manifold, Studies in Analysis, Adv. Math. (Supplementary) Academic Press, New York (1979), 233-239.

882. Suominen, K. Duality for coherent sheaves on an analytic manifold, Ann. Acad. Sci. Fenn. (A) 424(1968), 1-17.

883. Suwa, T. On hyperelliptic surfaces, J. Fac. Sci. Univ. Tokyo 16(1970), 469-476.

884. Suwa, T. Stratification of local moduli spaces of Hirzebruch manifolds, Rice Univ. Studies 59(1973), 129-146.

885. Suwa, T. Deformation of holomorphic Seifert fibre space, Invent. Math. 51(1979), 77-102.

886. Swarup, G. A. Homeomorphisms of 3-manifolds, Topology, 16(1977), 119-130.

887. Swinnerton-Dyer, H. R. F. An enumeration of all varieties of degree 4, Am. J. Math. 95(1973), 403-418.

888. Swinnerton-Dyer, H. R. F. Analytic theory of abelian varieties, Camb. Univ. Press (1974).

889. Tamura, I. Every odd dimensional homotopy sphere has a foliation of codimension one, Comm. Math. Helv. 47(1972), 164-170.

890. Tankev, S. G. On a global theory of moduli of surfaces of general type, Izv. Akad. Nauk SSSR 36(1972), 1200-1206.

891. Tanno, S. The spectrum of the Laplacian for 1-forms, Proc. AMS 45(1974), 125-129.

892. Tate, J. Genus change in inseparable extension of function fields, Proc. AMS 3(1952), 400

893. Tate, J. Homology of Noetherian rings and local rings, Ill. J. Math. 1(1957), 14-27.

894. Tate, J. 1974 Field Medal I, An algebraic geometer, Science (USA) October (1974).

895. Teichmuller, O. Bestimmung der extremalen quasikonformen Abbildungen bei geschlorsen orientierten Riemannschen Flachen, Preuss Akad. Ber. 4(1943).

896. Thimm, W. Meromorphe Abbildungen von Riemannschen Bereichen, Math. Z. 60(1954), 435-457.

897. Thom,R. Espaces fibres en spheres et curres de Steenrod, Ann.Sci.Ecole Norm.Sup. (3) 69(1952),109-182.

898. Thom,R. Quelques properties globales des varietes differentiables, Comm.Math.Helv. 28(1954),17-86.

899. Thom,R. Des varietes combinato aux verietes differentiables, Proc.Int.Congr.Math.Edinburgh (1958),Camb.Univ. Press (1970).

900. Thom,R. Les classes characteristiques de Pontrjagin des varietes triangulees, Symp.Intern.ToplAlg.(1956), Universidad de Mexico (1958),54-67.

901. Thom,R. Structural Stability and Morphogenesis,Benjamin, (1975).

902. Thom,R. On singularities of foliations,Tokyo (1973),Tokyo Univ.Press (1975),171-173.

903. Thomas,A.D. Zeta functions,An introduction to algebraic geometry, Research Notes in Mathematics,Pitman Publishing (1977).

904. Thomas,E. Fields of tangent 2-planes on one dimensional manifolds,Ann.Math. 86(1967),349-361.

905. Thomas,E. Fields of tangent k-planes on manifolds,Invent.Math. 3(1967),337-347.

906. Thomas,E. Complex structures on real vector bundles, Am.J.Math. 89(1967),887-908.

907. Thomas,E. The index of a tangent 2-field,Comm.Math.Helv. 42(1967),86-110.

908. Thomas,E. Vector fields on manifolds,Bull,Am.Math.Soc. 75(1969),643-783.

909. Thomas,E.,Wall,C.T.C. The topological spherical space form problem I, Comp.Math.23(1971),101-114.

910. Thurston,W.P. The theory of foliations of codimension greater than 1, Comm.Math.Helv. 49(1974),214-231.

911. Thurston,W.P. A generalization of the Reeb stability theorem, Topology 13(1974),347-352.

912. Thurston,W.P. On the construction and classification of foliations, Proc.Intern.Cong.Vancouver,Vol.1(1974/75),547-549.

913. Thurston,W.P. Existence of codimension one foliation, Ann.Math.(2) 104(1976),249-268.

914. Tits,J. Sur certaines classes d'espaces homogeneous de grupes de Lie,Mem.de l'Acad.Belgique,Brussels (1955)

915. Tjurin, A. N. On the deformation of complex structures of algebraic varieties, Soviet Math. Dokl. 4(1963), 1576-1574.

916. Tjurin, A. N. Locally semi-universal flat deformation of isolated singularities of complex spaces, Math. USSR Izv. 3 (1969), 967-99.

917. Tjurin, A. N. The Fano surface of a non singular cubic in $P^4$, Izv. Akad. Nauk SSSR Mat. 34(1970), 1200-1208.

918. Tjurin, A. N. The geometry of Fano surface of a nonsingular cubic F in $P^4$ and Tonelli's theorem for Fano surfaces and cubics, Izv. Akad. Nauk SSSR Ser. Mat. 35(1971), 498-529.

919. Tjurin, A. N. Five lectures on three - dimensional varieties, Russian Math. Surveys 27(1972), 1-53.

920. Tjurin, A. N. The geometry of the moduli of vector bundles, Russian Math. Surveys 29(1974), 57-88.

921. Todd, J. A. Canonical systems on algebraic varieties. Bol. Soc. Mat. Mexicana 2(1957), 26-44.

922. Trang, Le Dung, Ramanujam C. P. The invariance of Milnor number implies the invariance of topological type, Amer. J. Math. 98(1)(1976), 67-78.

923. Trang, Le Dung The geometry of the monodromy theorem, in C. P. Ramanujam, a Tribute (K. G. Ramanathan Ed.) Tata Institute, Bombay (1978), 157-173.

924. Trautmann, G, Ein Kontinuitatssatz fur die Fortesetgung Koharenter analytischen garben, Arch. Math. 18(1967), 186-196.

925. Trautmann, G. Deformations of sheaves and bundles, Springer Lecture Notes 683(1978), 29-41.

926. Trautman, E. Deformations of coherent analytic sheaves with isolated singularities, Proc. Symp. Pure Math. AMS (1975), 85-89.

927. Tsagas, G. On cohomology ring of a pinched Riemannian manifold, Math. Ann. 185(1970), 55-88.

928. Tsagas, G., Kockings, L. The geometry of the Laplace operator on exterior 2-forms on a compact Riemannian manifold, Proc. AMS 73(1979), 109-116.

929. Tyrrell, The Enriques threefold, Proc. Cam. Phil. Soc. 57(1961), 897-898.

930. Ueda, T. Compactification of $C^* \times C^*$ and $(C^*)^2$, Tohoku. Math. J. 31(1979), 81-90.

931. Ueno, K. Classification of algebraic varieties I and II, Comp. Math. 27(1973), 277-342.

932. Ueno, K. Classification Theory of Algebraic Varieties and Compact Complex Spaces, Lecture Notes 439, Springer Verlag (1975).

933. Ueno, K. On the pluricanonical systems of algebraic manifolds, Math. Ann. 216(1975), 173-179.

934. Varadarajan, V.S. Lie-Groups, Lie Algebras and their Representations, Prentice Hall (1974).

935. Ven, A Van De, Chern numbers of complex and almost complex manifolds, Proc. Nat. Acad. Sci. USA 55(1966), 1624-1627.

936. Ven, A Van De, On the Chern numbers of surfaces of general type, Invent. Math. 36(1976), 285-293.

937. Ven, A Van De, Some recent results on surfaces of general type, Seminar Bourbaki 76/77, Expose 500, Lecture Notes 677, Springer Verlag (1979).

938. Ven, A Van De, On the Enriques classification of algebraic surfaces, Seminar Bourbaki 76/77, Expose 506, Lecture Notes, 677, Springer Verlag (1979).

939. Ven, A Van De., Geer, Van Der. On the minimality of certain Hilbert modular surfaces; Complex Analysis and Algebraic Geometry ( In honour of Kodaira ), Iwanami Shoten, Tokyo (1977), 137-150.

940. Viehweg, E. Canonical divisors and the additivity of Kodaira dimension for morphisms of relative dimension one, Comp. Math. 35(1977), 197-223.

941. Vitter, A. L. Affine structures on compact complex manifolds, Invent. Math. 17(1972), 233-244.

942. Wahl, J. M. Deformations of plane curves with nodes and cusps, Am. J. Math. 96(4) (1974), 529-577.

943. Walker, R. J. Algebraic Curves, Princeton Univ. Press (1950), Dover Reprint 1962.

944. Wall, C. T. C. Classification Problem in Differential Topology I, Topology 2(1963), 253-261; ibid 2(1963), 263-272; III-Applications to special cases, Topology 3(1965), 291-304; IV-Classifications of (S-1) -connected (2s+1)-manifolds, Topology 6(1967), 273-296; V, Invent. Math. 1(1966), 355-374.

945. Wall, C. T. C. Diffeomorphisms of 4-manifolds, J. Lond. Math. Soc. 39(1964), 131-140.

946. Wall,C.T.C. On simply connected 4-manifolds,J. Lond. Math. Soc. 39(1964),141-149.

947. Wall,C.T.C. Surgery on non simply connected manifolds,Ann. Math. 84(1966),217-276.

948. Wall,C.T.C. Homeomorphisms and diffeomorphisms classification of manifolds,Proc. Int. Congr. Math. Moscow (1966). 450-460.

949. Wall,C.T.C. On Poincare complexes,Ann. Math. 86(1967),213-245.

950. Wall,C.T.C. Surgery on Compact Manifolds, Academic Press, New York (1970).

951. Wang.H.C. Closed manifolds with homogeneous complex structures, Amer. J. Math. 76(1954),1-32.

952. Wavrik,J.J. Obstructions to the existence of a space of moduli, Global Analysis,Princeton Univ. Press (1969).

953. Wavrik,J.J. A theorem of completeness for families of complex analytic spaces, Trans. Am. Math. Soc. 163(1971), 147-155.

954. Wavrik,J.J. A theorem of solutions of analytic equations with applications to deformations of complex structures, Math. Ann. 216(1975),127-142.

955. Weil,A. Foundations of algebraic geometry, An. Math. Soc. Coll. Publ. 29(1946),revised and enlarged (1962).

956. Weil,A. Varietes abeliennes et courbes algebrique, Hermann Paris (1948).

957. Weil,A. On Picard Varieties,Am. J. Math. 74(1952),865-894.

958. Weil,A. Abstract versus classical algebraic geometry, Proc. Int. Congr. Math. (1954),Vol. III,550-558.

959. Weil,A. Introduction a l'etude des varites Kahleriennes, Hermann,Paris (1958).

960. Weil,A. Sur les modules des surfaces le Riemann, Seminaire Bourbaki (1958).

961. Wells,R.O. Jr. Differential Analysis on Complex Manifolds,Prentice Hall (1973).

962. Wells,R.O. Jr. Complex manifolds and mathematical physics, Bull. AMS New Series 1(1979),296-336.

963. Weyl,H. Die Idee der Riemannschen Flache, Teubner,Berlin, (1913),(Available in English).

964. Weyl,H. The Classical Groups,Princeton Univ. Press

965. Weyl, H. David Hilbert and his Mathematical Work, Bull. Am. Math. Soc. 50(1944), 612-654.

966. Weyl, H. Ramifications, old and new of the eigenvalue problem, Bull. Am. Math. Soc. 56(1950), 115-139.

967. Whitehead, G. W. Fibre spaces and the Eilenberg homology groups, Proc. Nat. Acad. Sci. USA 38(1952), 426-430.

968. Whitehead, G. W. Recent advances in homotopy theory, Conf. Board of Math. Sci. Regional Conf. In Math. AMS (1970).

969. Whitehead, G. W. Elements of Homotopy Theory, Graduate Text in Math. 61, Springer Verlag (1978).

970. Whitehead, G. W. On $C^1$ complexes, Ann. Math. 41(1940), 809-832.

971. Whitehead, J. H. C. On simply connected 4 dimensional polyhedra, Comm. Math. Helv. (1949), 48-92.

972. Whitehead, J. H. C. Combinatorial topology, B. A. M. S (1949), 213-245, 453-496.

973. Whitney, H. Differentiable manifolds, Ann. Math. 37(1935), 645-680.

974. Whitney, H. Complex Analytic Varieties, Addison-Wesley (1972).

975. Wilder, R. L. Topology of manifolds, Am. Math. Soc. Coll. Publ. (1949).

976. Wilson, P. M. H. The behaviour of the plurigenera of surfaces under algebraic smooth deformations, Invent. Math. 47(1978), 289-299.

977. Wolffhardt, K. Variation of a complex structure in a point, Am. J. Math. 90(1968), 553-567.

978. Wong, B. Singularities of spaces of flat bundles over complex manifolds, Trans. Am. Math. Soc. 231(1977), 451

979. Wood, J. Foliations of codimension one, Bull, Am. Math. Soc. 76(1970), 1107-1111.

980. Wright, M. The Kobayashi pseudometric on algebraic manifolds of general type and deformation of complex manifolds, Trans. Am. Math. Soc. 232(1977), 357-370.

981. Wu, W. T. Sur les classes characteristiques de structures fibrees spheriques, Actualities Sci. Ind. Hermann, (1952), 5-89.

982. Wu. W. T. Sur les espaces fibres, Actualities Sci. (1952), 1183

983. Yang, C. N., Mills, R. L. Physics Reviews 96(1954), 191

984. Yang, C. T. Hilbers's fifth problem and related problems on transformation groups, Proc. Symp. Pure Math. 28(1976), 142-146.

985. Yau, S. T. On the curvature of compact Hermitian manifolds, Invent. Math. 25(1974), 213 - 239.

986. Yau, S. T. Intrinsic measures on compact complex manifolds, Math. Ann. 2-2(1975), 317-329.

987. Yau, S. T. On Calabi's conjecture and some new results in algebraic geometry, Nat. Acad. Sci. USA 74(1977), 1798-1799.

988. Yau, S. T. Parallelisable manifolds without complex structures, Topology 15(1977), 51-53.

989. Yau, S. T. On the Ricci curvature of a compact Kahler manifold and the complex Monge-Ampere equation I, Comm. Pure. Appl. Math. Vol. XXXI (1978), 339-411.

990. Yoshinagu, E. On flat families and complete families of analytic spaces onto complex manifold, Math. Z. 116(1970), 258-263.

991. Zappa, G. Sull esislenza sorpa le superficie algebrische di sistime continui completi infuile, la curi curva generica e serie characteristica incompleta, Pol. Acad. Sci. Acta 9(1945), 91-93.

992. Zariski, O. The fundamental ideas of algebraic geometry, Proc. Int. Congr. Math. (1950), Vol. II, 77-89.

993. Zariski, O. Complete linear systems on normal varieties and a generalisation of a Lemma of Enriques-Severi, Ann. of Math. 55(1952), 552-592.

994. Zariski, O. On Castelnuovo's criterion for rationality $p_a = p_2 = 0$ of an algebraic surface, Illinois J. Math. 2(1958), 303-315.

995. Zariski, O. Introduction to the problem of minimal models in the theory of algebraic surfaces, Publ. Math. Soc. Jap. 4 (1958).

996. Zariski, O. Studies in equisingularity I, Am. Math. 87(1965), 507-536; II, ibid 972-1006; III, Am. J. Math. 90(1968), 961-1023.

997. Zariski, O. Algebraic Surfaces, 2nd Ed. Springer Verlag (1971).

998. Zariski, O. Some open questions in the theory of singularities, Bull. Am. Math. Soc. 77(1971), 481-491.

999. Zeeman, E. C. The generalised Poincare conjecture, Bull. Am. Math. Soc. 67(1961), 270

1000. Zeeman, E. C. Seminar on combinatorial topology, IHES, Paris (1963).